The Grassroots of a Green Revolution

The Grassroots of a Green Revolution
Polling America on the Environment

Deborah Lynn Guber

The MIT Press
Cambridge, Massachusetts
London, England

This book was set in Sabon on 3B2 by Asco Typesetters, Hong Kong. Printed on recycled paper and bound in the United States of America.

Library of Congress Cataloging-in-Publication Data

Guber, Deborah Lynn.
The grassroots of green revolution : polling America on the environment / Deborah Lynn Guber.
p. cm.
Includes bibliographical references and index.
978-0-262-07238-0 (hc: alk. paper) — 978-0-262-57160-9 (pbk: alk. paper)
1. Environmentalism—United States—Public opinion—History—20th century. 2. Environmental protection—United States—Public opinion—History—20th century. 3. Public opinion—United States. I. Title.
GE197 .G84 2003
363.7'0525—dc21   2002070329

10 9 8 7 6 5 4 3 2

For my parents,
for reasons more than words can say.

And for my nephew Ethan,
who reminds me
so much of his mother.

Contents

Acknowledgments

As with almost any project that demands a level of dedication best measured in years, the evolution of this book has followed a long and circuitous path. While I was a graduate student at Yale University in the mid-1990s, a project that was conceived of initially as a seminar paper quickly grew into a series of conference presentations and finally into a doctoral dissertation. With the invaluable input of academic advisors, panel discussants, anonymous manuscript referees, faculty colleagues, and students, that work eventually matured into a series of articles published in scholarly journals, which were transformed yet again with fresh data and better ideas into what appears on paper here. Any remaining errors are, of course, mine alone.

I laud the efforts of Clay Morgan, Sara Meirowitz, and Deborah Cantor-Adams at the MIT Press. Their collective and keen attention to detail has improved this manuscript in a myriad of ways. I am indebted to Riley E. Dunlap and Christopher J. Bosso for their detailed comments and enthusiastic suggestions. Without their guidance this project would be far less satisfying. I am grateful also to my colleagues at the University of Vermont for creating an atmosphere in which I have felt challenged intellectually and welcomed personally. In particular, I would like to thank Candace Smith and Carol Tank-Day for their gracious administrative support; Howard Ball, Phillip Cooper, and Robert Taylor for their willingness to read my manuscript while it was in various stages of disorder; Caroline Beer, for her methodological and statistical savvy; Frank Bryan for his lightning wit and boundless wisdom; Gregory Gause for countless hours of advice, much of which (to my discredit) I did not

follow; and Robert Kaufman, who likes to think of himself as my tor-
mentor but who is in secret one of the nicest men I know.

I would also like to recognize those teachers who have inspired me
throughout the years to think hard and work harder and who have
been mentors in the truest sense of the word, including John Salamone,
Donald C. Baumer, Roger T. Kaufman, Catherine Rudder, Donald P.
Green, John P. Wargo, and Sarah McMahon. And finally, I am grateful
beyond words to family and friends who persevered alongside me in this
journey, most especially to those two who did not live to see its end. By
supporting me always with great forbearance and a healthy sense of
humor, this book is in many ways yours as well as mine.

The Grassroots of a Green Revolution

Introduction
Environmental Concern and the Politics of Consensus

In the main, the notion of consensus has sprung from the inventive minds of theorists untainted by acquaintance with mass attitudes.
—V. O. Key, Jr.[1]

In the generation that has passed since the first Earth Day in 1970, environmentalism has become woven into the fabric of American life. Concern for environmental quality has spawned extraordinary changes in how we think, work, and recreate, in what we buy, and how we govern. Words like "ecology," "acid rain" and "global warming" have become common in the lexicon of our language, sorting newspapers, bottles, and cans into bins a daily household ritual. Indeed, society has been so altered by environmentalism it is easy to overlook the distance of thirty years, backwards in time to the nascent social movement that was, and forward again to the mature evolution of science and law it has become.

Today, environmentalism is a part of our popular culture and a reflection of modern sensibility, reinforced by what we read, see, and hear in books and magazines, on television, and at the movie theater. Our collective consciousness has been raised by the writings of those Charles Rubin calls "popularizers," including Rachel Carson, Barry Commoner, Paul Ehrlich, and the Club of Rome.[2] The authors of children's books promote environmental themes,[3] product placements on popular television shows send subtle environmental cues,[4] motion pictures like *Erin Brockovich* and *A Civil Action* remind us of what we fear and who we should blame.[5] As one newspaper columnist wryly observed, "Motherhood and apple pie, baseball and the flag—all may be subjects of controversy. But the environment is almost beyond debate these days."[6]

V. O. Key, arguably the founding father of modern scholarship on public opinion, had a word for such a pattern. He called it (simply enough) *consensus*, although he also warned that it was a "nebulous term" with many meanings, full of uncertainty, with the potential to obscure as much as it illuminates.[7] For Key, however, identifying the presence or absence of consensus was only the first step, and a relatively straightforward one at that, at least in comparison to the true task of determining its impact on the political system. While he found that under some conditions overwhelming public agreement on an issue performed a "decisive" function, forcing change in existing policies and programs, at other times its influence was less pronounced. In some cases, he speculated, the function of consensus was merely "supportive" of government; on other occasions it was surprisingly "permissive," allowing government to act largely without fear of popular dissent or electoral reprisal. In some ways, too, the appearance of consensus might even be "contrived" to suit political needs and purposes. The product of many different things, all consensus was not created equal.

Within that context, acknowledging that public attitudes on the environment approach consensus is clearly not enough. Given the need to translate words into action, to use the word *consensus* (even judiciously) reveals little about its long-term political impact, at least from the bottom up. On that note, confusion abounds. Since 1970, surveys have demonstrated widespread public concern for a growing list of environmental problems, including air and water pollution, nuclear power, energy conservation, deforestation, and urban sprawl. Public opinion polls likewise show that the environmental movement has earned the sympathetic support of a large majority of Americans, many of whom claim the label *environmentalist* as their own. But what do the numbers that underlie such research ultimately mean? Is environmentalism a shallow consensus likely to soften in the face of ambivalence, as American voters and consumers experience the costs of reform firsthand? Is it an enduring social concern or a fleeting political fashion subject to the nature-of-the-times as the economy shifts from prosperity into recession? Does growing support for environmental protection indicate a fundamental shift in American values and lifestyles, or is it merely an uncontroversial "motherhood" issue that engenders automatic support without personal commitment or lasting political consequences?

Satisfying answers are not immediately apparent—to environmentalists committed to translating public sympathy into political currency, politicians held responsible for answering constituent demands, scientists and policymakers frustrated by public misperceptions and misdirected fears, marketing executives charged with identifying and satisfying environmentally concerned consumers, or academics who continue to disagree. No one quite seems to know what to make of the environment.

The State of the Movement

If the celebrations marking the thirtieth anniversary of Earth Day are any indication, the news for environmentalists is both good and bad. Environmental causes resonate with most Americans, to be sure—83 percent of those polled in an April 2000 Gallup poll readily agreed with the broadest goals of the environmental movement—but when asked to rate their own commitment to the cause, just 16 percent said they were "active participants," while more than half admitted they were sympathetic but uninvolved.[10] These statistics run parallel with broader trends in declining membership among some national environmental organizations since 1990.[9]

Moreover, when asked to rate the seriousness of various problems on the national agenda in a comparative sense, respondents ranked the environment well behind other issues and concerns, including drug use, crime and violence, health care, and homelessness. If Americans are environmentalists, Hal Rothman suspects they are "half-hearted" ones, at best, and are unwilling to face difficult choices and altered lifestyles. In fact, he writes in *The Greening of a Nation?* (1998) that the contemporary environmental movement has become, ironically, a "victim of its own successes." By finding appeal in popular culture, he warns, it has become too easy, to pay "lip service to the concepts of environmentalism without engaging in the behaviors necessary to turn concepts into action."[10]

Motivating and sustaining the political activism of average Americans has been an uphill battle for the environmental movement from the start. Communicating the complex nature of environmental destruction to a lay public that is not expert in science and technology required "popularizers" like Rachel Carson and Barry Commoner to resort to stories

and simplifications. According to Charles Rubin, the unintended consequence of this approach was to create a "public taste" for grand tales of
ecological disaster. Books like Carson's *Silent Spring* (1962) and Commoner's *The Closing Circle* (1971), he says, became

> the intellectual equivalent of a gothic romance, with a large cast of characters,
> involuted relationships, and a lurking menace. But the public's ability to appre-
> ciate the delicate balances and interrelationships of political and social structures
> has undergone a corresponding debasement, evident in rampant sloganeering,
> shameless emotionalism, and mindless panic and pessimism whenever "what is
> wrong with our society" comes under discussion. In this realm, only the crudest
> morality tales satisfy. Carson and Commoner have alerted us to matters that may
> well demand our attention. But they have done so at the cost of our ability to
> give that attention in a thoughtful way.[11]

In short, by downplaying environmental progress and by using exaggerated doomsday warnings to motivate public awareness and concern, the
environmental movement has sacrificed its own credibility by giving in to
the politics of chicken little.[12]

It is a common complaint among recent critics of the environmental
movement, one voiced by Mark Dowie in *Losing Ground* (1995) and
even more forcefully by Gregg Easterbrook in *A Moment on the Earth*
(1995) and Bjorn Lomborg in *The Skeptical Environmentalist* (2001),
but it is, in many ways, difficult to deny.[13] As David Brower, former
executive director of the Sierra Club, remarked in an Earth Day 2000
interview: "All I've done in my career is slow the rate at which things get
worse. Basically, that's all the environmental movement has done during
the past thirty years."[14] Or as Donella Meadows, of *Limits to Growth*
fame, recently wrote in her syndicated newspaper column: "If in the 30
Earth Day celebrations since 1970, the human population and economy
have become any more respectful of the Earth, the Earth hasn't noticed."
Ultimately, she too refuses to give in to the "die-hard optimists."[15]

In the end, the occasion marked by both Brower and Meadows—the
annual celebration of Earth Day—is a prime example of the fundamental
tension between popularity and ideology.[16] While environmentalists
continue to blame many of the earth's problems on rampant over-
consumption, organizers of recent Earth Day events were nonetheless
quick to offer T-shirts, tote bags, coffee mugs, solar calculators, and
hemp backpacks for sale to an appreciative audience, the gross sum of

which prompted some observers to complain that Earth Day had become little more than a commercial occasion, overrun by "vacuous celebrities."[17] In the past, too, others have noted the hypocrisy of allowing polluters the privilege of "greenwashing" their records by signing on as corporate sponsors of Earth Day events.[18] Following Earth Day 1990 in New York City, during which two hundred thousand people gathered in Central Park, creating almost forty-five tons of garbage, Rothman complained, "It is entirely possible that the planet might have been better off if they had just stayed home."[19]

The Political Arena

If environmentalists have struggled in a public relations war, politicians have not fared much better in gauging their voting leeway on an issue marked by a combination of chronic low salience and high issue support.[20] It was low salience that led the Reagan administration in the early 1980s to assume that the public would be willing to back away from strict environmental regulations to revitalize the economy. Yet environmental concern and public furor soon galvanized over political appointees such as James Watt and Anne Burford, eventually forcing Reagan to change course by substituting administrators more sensitive to environmental causes.

After a sustained period of public outrage, however, the environment failed to materialize as a significant issue during Reagan's campaign for reelection in 1984. Looking at his meager environmental record, his unrivaled attention to deregulation and economic growth, and the lingering controversy over his appointments of Watt and Burford, many environmentalists believed the Reagan record would lead to political liability at the ballot box.[21] Even though polls suggested that voters were both aware and disapproving of Reagan's record on the environment, in the end the issue had little effect on his political success.[22]

Nearly every politician on the national scene has had to contend with the public's mixed signals on the environment ever since. Campaigning for president in 1988, Republican candidate George Bush pledged to be an "environmental president" in the grand tradition of Teddy Roosevelt, and yet during an economic slump just four years later Bush seemed to reverse course, insisting instead that it was "time to put people ahead of owls."[23] Even Al Gore, whose strong environmental convictions made him an attractive running mate for Democrats in 1992, has since turned cautious, tempering his environmental views during his own quest for the White House during the 2000 presidential campaign.[24] As *The Economist* noted: "Despite his talk of bold measures and radical solutions ... what he offers is virtue without sacrifice. As a political programme this is hard to beat," but it does little to help the environment.[25]

Gore's political dithering notwithstanding, Republicans have traditionally faced the most significant ideological challenge on environmental issues.[26] An overwhelming number of Americans favor environmental protection through government intervention in the market economy, a principle resisted by fiscal conservatives. At the same time, however, "a strong backlash" has developed against environmental regulations that are viewed as "intrusive, bureaucratic, and overly protective," opening to the door to an odd triangulation.[27] Following a Republican sweep in Congress in 1994, the new partisan majority was quick to propose cuts in the budget for the U.S. Environmental Protection Agency and amendments that would have weakened the Clean Water Act and the Endangered Species Act. Republicans moved to safeguard private property rights, close parts of the national park system, and increase oil drilling in the Arctic National Wildlife Refuge. Yet under pressure not long after, House Speaker Newt Gingrich (Republican-Georgia) sought a rapid public relations retreat, admitting that Republican proposals were in fact "strategically out of position on the environment."[28]

Understanding Risk

Confusion and frustration over the political impact of environmental concern extends well beyond the strategies used by candidates to win elections. Given the need to justify federal regulatory decisions on the basis of science, it also includes the growing field of risk communication. "Any one of us might be harmed by almost anything," writes Stephen Breyer in *Breaking the Vicious Circle* (1993)—"a rotten apple, a broken sidewalk, an untied shoelace, a splash of grapefruit juice, a dishonest lawyer."[29] It is, of course, the responsibility of government (and by extension the experts they employ) to decide which of those multitude of risks merit regulation and which do not. According to Breyer, however, several factors intrude on our ability to rank those risks wisely, not the least of which involve public misperceptions and exaggerated fears, factors he says "impede rational understanding."[30]

Scholars have long recognized that public perceptions of risk often collide with what experts judge to be objective probabilities of harm. While scientists cite motor vehicle use, smoking, and alcohol consumption as three of the riskiest activities of modern life, lay people instead believe nuclear power to top that list, often underestimating fatalities caused by less "dramatic" accidents and diseases, while overemphasizing the magnitude of danger to be found in new technologies.[31]

The same gap between the concerns of average Americans and those of policymakers can be found across a wide range of environmental problems. While environmentalists promote global warming, ozone depletion, deforestation, and loss of habitat for endangered wildlife, respondents in public opinion polls are far more likely to worry about mainstay issues like air and water pollution. As Jonathan Rauch, a columnist for the *National Journal*, recently observed: "The public's priorities almost perfectly invert the environmental movement's priorities. Perversely, the aspirations of Gore-era environmentalism are now blocked by the public's commitment to Nixon-era environmentalism."[32]

In the end, this disjunction has significant consequences for policymakers. According to Breyer, without better risk communication from the top down, public attitudes toward environmental risks often remain stubborn and unyielding, warping political priorities and pressuring scarce resources of time and money into all the wrong places.[33]

In Search of the Green Consumer

While environmentalists measure support for the environment using membership rolls, politicians rely on votes, and scientific experts depend on public trust, corporate America has hoped to translate rising environmental concern into increased sales in the marketplace. Driven by the desire of American businesses to exploit consumer demand, great effort over the past two decades has been spent on identifying and targeting the "ecologically concerned consumer." Despite what some environmentalists feel is an obvious oxymoron, today's eco-market offers an unusually diverse collection of products and services—from phosphate-free detergents and recycled paper products, to electric cars, environmentally responsible mutual funds, solar mosquito repellents, and herbal flea collars.[34]

The results have been decidedly mixed, despite the involvement of visible corporate giants like Proctor & Gamble, Wal-Mart, and McDonald's. While most studies find that deep commitment to the environment is concentrated in the hands of a privileged few—ranging from 5 to 25 percent of the U.S. population, depending on the stringency of the criteria used[35]—public willingness to purchase certain environmentally friendly products runs surprisingly deep, even at slightly higher cost. A 1992 survey, for example, found that nearly three-quarters of consumers were "at least sometimes" influenced by environmental claims in the marketplace, and most appeared willing to pay at least 5 percent more for products known to be environmentally safe.[36]

While some have welcomed environmentalism as "the political, economic, and social trend of the '90s," others suspect that when pressed Americans fail to put their money where their mouth is.[37] For example, despite evidence that many Americans prefer and indeed are willing (when asked in surveys) to pay a premium price for environmentally safe products, Universal Product Code (UPC) scanner data and panel studies that trace actual buying behavior often paint a more lackluster picture. The "Study of Media and Markets" by Simmons Market Research Bureau finds that the products that consumers purchase "most often" (such as aerosol sprays and radial tires) are frequently at odds with their environmental preferences and their stated willingness to purchase substitutes.[38] As one observer put it, environmental concern alone does not always "make the cash register ring."[39]

Getting It Right

The four brief vignettes presented above crisscross a wide range of experiences and disciplines, and yet all illustrate the importance of achieving a better understanding of public opinion on environmental issues.

First and foremost, to study the success of the environmental movement from the bottom up means confronting a movement that is, paradoxically, both strong and weak. Widespread and well-meaning public concern at the grassroots level has become one of the most impressive findings in recent survey research, and yet it is a resource environmentalists have been unable (or unwilling) to capitalize on fully. On the one hand, the discomfort some environmentalists feel toward the ideological impurity of their rank and file leads, at times, to evangelical and exclusionary rhetoric. On the other, motivating public outrage by instilling fear seems a short-sighted solution to a long-range problem, one that ultimately risks the credibility of the movement itself and leaves sympathetic supporters feeling demoralized about the insignificance of their efforts.

As Matt Ridley and Bobbi Low remind us, "At the center of all environmentalism lies a problem: whether to appeal to the heart or to the head—whether to urge people to make sacrifices on behalf of the planet or to accept that they will not, and instead rig the economic choices so that they find it rational" to behave responsibly in any event.[40] Environmentalists, they believe, will never achieve their goals simply by occupying the moral high ground. If they are to motivate and mobilize latent support among average Americans, environmentalists need to become more proficient at communicating with the public by first understanding the root of its concerns.[41]

Second, politicians unsure of whether to court, fear, or ignore the "green" vote would also do well to scrutinize the factors that underlie public attitudes toward the environment. As politicians from both sides of the political fence learn to embrace the environmental issue as their own, the ability to harness latent public concern seems likely to become an increasingly important political and electoral skill, creating, in Christopher Bosso's words, "opportunities for leadership that may or may not be exploited."[42] While motivating environmentally conscious voters at the ballot box may not be easy or fail-safe, it is surely a strategy worth research and investigation, especially in close races, where victory is won at the margins.

On other political fronts, as the arena of science and policymaking becomes ever more democratic, the need for effective risk communication between experts and the lay public likewise becomes critical. Faced with a public that is unable to distinguish good science from "junk science," the political process designed to mediate between both sources of information is frequently plagued by uncertainty and distrust. As Breyer notes, "To change public reaction, one would either have to institute widespread public education in risk analysis or generate greater public trust in some particular group of experts or the institutions that employ them."[43] But to achieve either goal requires a firm understanding of how attitudes develop and how they adapt in response to new knowledge and information.

Finally, understanding environmental attitudes as a stimulus to marketplace behavior is clearly important to Madison Avenue, where advertisers struggle to hone and refine their environmental messages on product labels. With a number of visible false starts and missteps, corporations looking to expand their markets further need to understand more than narrow consumer preferences. Given that many environmental issues span the chasm between public and private, blurring lines of distinction between citizen and consumer, political attitudes toward the environment will likely become important in understanding individual economic decisions.

The Academic Divide

Given increased attention and a continued state of controversy, it might seem reasonable to assume that in the academic disciplines considerable progress already has been made toward understanding the origins and importance of public opinion on environmental issues. "Instead," as Kent Van Liere and Riley Dunlap deplore, even on relatively simple questions, such as the social and demographic bases of environmental concern, "one finds considerable dissensus with respect to both the evidence itself and its interpretation."[44]

For instance, in *Progress and Privilege: America in the Age of Environmentalism* (1982), William Tucker insists that "At heart, environmentalism favors the affluent over the poor, the haves over the have-nots."[45] Yet Mark Sagoff disagrees, despite a lingering misperception, and argues that environmentalism "serves as a common rallying ground for groups usually thought to be at odds with one another: educated professionals and the lower middle class; affluent suburbanites and inhabitants of small towns in the American heartland."[46] As such, he adds, environmentalism represents an entirely new breed of populism. Strengthened by cross-cutting cleavages, it is a movement centered not on elitist principles but rather around a sense of community and the integrity of place.

Other disagreements continue to be fought in the pages of scholarly books and journals as well. A few examples illustrate the point:

• David Gelernter contends that "There is no such thing on the political scene as an 'anti-environmentalist,' no cogent intellectual position by that name,"[47] and yet Jacqueline Vaughn Switzer charts a growing ideological challenge from the grassroots—a resentment and distrust of political and environmental leaders that she believes amounts to an effective "green backlash."[48]

• Anthropologists Willett Kempton, James Boster, and Jennifer Hartley believe that environmental activists "have more or less the same beliefs as other Americans," despite media stereotypes to the contrary.[49] Richard Ellis and Fred Thompson, however, disagree. They find that "Americans do not behave more like environmental activists because culturally they are quite unlike them." To suppose otherwise, they say, is to "miss the rival value systems that undergird environmental policy debates."[50]

• Laura Lake comments on the "depth and longevity of the environmental mandate,"[51] and yet polling expert William Schneider speculates that "because the consensus is so broad, it is not likely to have much impact on politics."[52] Like Lake, Riley Dunlap insists that success on key environmental initiatives and referendums proves that Americans are willing to take a stand on environmental issues, especially when elected representatives fail to do so,[53] but in *Earth Rising* (2000) Philip Shabecoff notes that the environment has been a "minor, rather ineffectual player in the electoral process."[54]

In short, despite a growing field of academic expertise spanning politics, economics and the sciences, few uncontested answers have emerged to a ever growing laundry list of questions.

An Overview of This Book

"To speak with precision of public opinion," said V. O. Key, is a task not unlike coming to grips with the Holy Ghost."[55] This book faces that challenge head on—by exploring the ironies, myths, inconsistencies, and tensions that characterize public thinking on environmental issues. To reach that goal, the essays that follow use original analyses of public opinion polls to break the problem into component parts—into those descriptive pieces Key referred to as "properties" or "qualities" of public opinion. Then, much like a jigsaw puzzle, the conclusion of this book reassembles those pieces back into a coherent whole, one that ultimately weighs the significance of environmental concern in the arena of U.S. politics and policy and provides some pragmatic advice for decision makers. It proceeds in the following order.

Direction

Chapter 1 opens with the deceptively simple task of characterizing the direction of public attitudes on environmental issues. Surely we might expect clear answers and little variance here, since a preponderance of surveys demonstrate that almost "everybody is an environmentalist these days."[56] But in the end, results are surprisingly complex, demonstrating that while environmental consensus exists on the idea of environmental protection, it does not always extend to the means used to achieve those goals. Conclusions here also reinforce the need to pay careful attention to potential biases embedded in survey questions and design, particularly in those polls sponsored by organizations with a vested interest in the outcome.

Strength

The direction of opinion is refined further by adding attitude strength into the mix in chapter 2. A crucial consideration to early scholars such as James Bryce and A. Lawrence Lowell, as well as to more modern academics like V. O. Key, Howard Schuman, and Stanley Presser, the question of intensity, says Key, clearly "puts us on the trail of a significant aspect of the place of opinion in the governing process."[57] Using measures of willingness-to-pay (WTP) as well as counterarguments and trade-offs that force respondents to consider the costs of achieving environmental goals relative to other priorities, survey data here largely confirm what scholars have suspected all along: Americans are quick to embrace environmental issues but reluctant to accept the consequences of their demands.

Stability

To probe the limits of popular support further, chapter 3 tracks the aggregate stability of environmental attitudes over time, especially in the face of economic recession and rising energy costs. Here survey data are used to explore the conventional wisdom that public enthusiasm for the environment is transient and that it responds to trends in the business cycle, rising policy costs, media attention, or simple boredom as new issues rise to complete for scarce public attention.[58]

Distribution

The distribution of public opinion on the environment among social and demographic groups is described in chapter 4, revisiting long-standing debates about the elitist character of the environmental movement and its key supporters.[59] Factors such as age, education, income, race, gender, and partisanship are all used in empirical models to explain variance in a variety of measures of environmental concern, ultimately with little effect. Making up in breadth what it lacks in depth, environmentalism on the surface appeals to nearly everyone.

Constraint

Taken together, do all of these pieces suggest (at least in nascent stages) the development of a fundamentally new social paradigm or belief system—one that might grow in commitment with time and patience? Chapter 5 draws attention to that issue by examining the consistency of environmental attitudes—what Key called "interrelations of opinion."[60] Scholars have long puzzled over low correlations between different environmental measures used in surveys, with some suggesting that public attitudes on the environment are rather crude, disconnected, and narrowly focused.[61] Yet after developing a more sophisticated model to test the dimensionality of environmental concern, results presented in this chapter largely demonstrate the opposite. Although notably lacking in knowledge and sophistication, public opinion on the environment is surprisingly consistent and "constrained."[62]

Behavior

Part II of the book moves one step further toward understanding the real impact of consensus by exploring the link between public opinion and political behavior. Chapter 6 examines the impact of environmental attitudes on American political parties and their candidates. That focus is expanded in chapter 7 to include other electoral arenas and political forums, including statewide referendums and initiatives. Finally, chapter 8 examines the impact of environmental concern on consumer decisions in the marketplace. That Americans prefer activism on the consumer front is surprising to scholars, perhaps, but this pattern of behavior reveals much about the factors that motivate citizens to act in an environmentally responsible way.

"In its most uncomplicated form," wrote Key, "'consensus' means an overwhelming public agreement upon a question of public policy." Yet, he added, it is also a "magic word" full of the uncertainty of interpretation.[63] With high issue support but low intensity of belief, with broad appeal but narrow participation, and with consistent opinions nonetheless ungrounded by clear knowledge and understanding, the long-term success of the U.S. environmental movement from the bottom up remains unclear. Growing environmental consensus may force elected officials from both sides of the political fence into action, but it might also allow political leaders a considerable degree of latitude in designing environmental policies free from a watchful public eye. Sympathetic public support may invest the environmental movement with valuable political currency, but an environmental consensus based on broad symbolism alone might prove to be shallow and manipulable in the long run, endangering the legitimacy and political base of environmentalists who place too much faith in public mandates and grassroots support.

Ultimately, the importance of these lingering, unresolved issues ensure that the legacy of American environmentalism will be determined not only by the success of its legislative record but also by its ability to persuade average citizens to change their voting patterns, buying habits, and lifestyles. As Pogo the Possum once said in a famous cartoon: "We have met the enemy and he is us." Whether or not those enemies have in fact become allies in the shadow of this "great environmental awakening" will be a matter debated throughout the coming pages.[64]

I

Attitudes

1

Direction: Do Americans Favor
Environmental Protection?

In 1997, with world leaders about to meet in Kyoto, Japan, for final negotiations on an accord to reduce emissions of greenhouse gases, the Pew Research Center for the People and the Press sponsored a poll to determine where American thinking on the topic stood. The results seemed remarkable: with a majority of those polled backing international standards, a columnist for the *Los Angeles Times* said the survey "left business lobbyists stunned and environmentalists gloating."[1] Even more significant was evidence that respondents were willing to accept higher gasoline prices as well—from five to twenty-five cents higher per gallon—in return for steps taken to counter the threat. In astonished response, the *Atlanta Journal and Constitution* joined a chorus in the news media in declaring that Americans were willing at long last "to reach into their pocketbook to help reduce global warming."[2]

Support for increasing gasoline prices notwithstanding, high levels of public support for environmental protection should come as no surprise. As former EPA administrator Carol Browner once put it, "I have yet to meet a member of the public who thinks their air is too clean or their water is too safe."[3] Yet while it may be tempting to diminish the significance and popularity of environmental concern by believing it represents little more than the political equivalent of "motherhood and apple pie" —a goal so infused with social desirability that it is nearly impossible to oppose—public opinion polls over the past thirty years have been adamant, demonstrating time and again that the environmental movement has earned the genuine sympathy of a large majority of Americans. Like the Pew Research Center poll on global warming, it is a conclusion well charted by newspaper headlines from around the country, including:

Environment Is a Big Concern for Californians, Poll Shows[4]

Voters Support Rules on Pollution; GOP Attacks Not Popular[5]

Polls Show Texans Wanting to Recycle[6]

Wilderness Expansion Backed; 80 Percent Favor Land Protection in Colorado Poll[7]

In response to such overwhelming numbers, the director of the White House Office of Management and Budget Richard Darman acknowledged in 1990 that, at heart, "we are *all* environmentalists. The President is an environmentalist, Republicans and Democrats are environmentalists. Jane Fonda and the National Association of Manufacturers, Magic Johnson and Danny DeVito, Candice Bergen and the Golden Girls, Bugs Bunny and the cast of *Cheers* are all environmentalists."[8] In this sense, to begin an analysis of public attitudes on the environment with an extended discussion of a single characteristic—the "direction" of opinion—might seem needless and simplistic, even a bit wasteful of time and effort. At least this much is apparent: Americans are worried about the health of the natural environment and want to see something done to protect it. The tenacity of those beliefs might be questioned, and the willingness of citizens to translate concern into commitment might be doubted, but it hardly seems controversial to suppose (under the best of intentions) that "everybody is an environmentalist these days."[9]

In reality, however, survey research on the environment is rarely indisputable, even on deceptively simple questions. High margins of support and strong expressions of concern can, at times, mask considerable disagreement surrounding the means used to achieve policy goals. Secondhand reports of poll results in the popular press can remove data results from their original context, allowing environmental concern to appear more impressive than it is. Finally, biased and misleading questions administered to unsuspecting respondents by groups with a vested interest in the outcome can so alter environmental responses that the results say more in the end about the mechanics of the poll itself than the subject it attends.[10]

Given that the foundation of this book is built piece by piece on evidence drawn from national public opinion polls, an initial note of warning about distinguishing good data and fair reporting from bad is clearly warranted. The purpose of this chapter is to confront each of these issues up front and early on and to identify (and avoid) red flags that have the potential to mislead even the most cautious reader. The close of this chapter then applies those lessons to a specific and detailed case, returning to the debate over global warming in the months leading up to the Kyoto Protocol, demonstrating that when it comes to the environment, public opinion is not always what it seems.

The Boundaries of Consensus

While V. O. Key was perhaps the first scholar to recognize the significance of opinion consensus in American politics, anthropologists Willett Kempton, James Boster, and Jennifer Hartley were the team that deepened its cultural meaning and applied the term in a vigorous manner to the environment. In *Environmental Values in American Culture* (1995), they argue that "Americans share a common set of environmental beliefs and values" and that on environmental issues there is, in fact, a single consensus with only "one set of culturally agreed upon answers."[11]

Their conclusions are based on an ambitious study combining open-ended interviews with a fixed-form questionnaire administered to a small group of respondents ($n = 142$). Divided into five diverse categories intended to represent a continuum of opinion, two groups were composed of self-avowed environmentalists (members of Earth First! and the Sierra Club), two represented occupations harmed by environmental legislation (dry cleaners and unemployed sawmill workers), and a final group was drawn from members of the general public in California.

While these groups might be expected to view the environment and human interactions with it in very different ways, Kempton and his colleagues find instead a surprising pattern of agreement. For instance, strong majorities in all five groups agreed that "there is a need to protect the environment because humans depend upon it," rejecting claims that environmental protection is either excessive or unnecessary.[12] Members of these groups also expressed an understanding of limited resources, fragile interdependencies in nature, and the dangers of disrupting natural processes, believing in the end that "it is more costly to fix problems than it is to prevent them in the first place."[13]

Despite the statistical limitations of their small sample, the conclusions reached by Kempton, Boster, and Hartley are hardly surprising. In fact, the shared values and beliefs they identify, which they say are "now closely tied to many other deep value systems in American culture," are those that would seem to underlie and explain recent trends in national polls on environmental topics.[14] For example,

• Of those surveyed in an April 2000 Gallup poll, 83 percent agreed with the goals of the environmental movement, although a majority feared that only "some progress" had been made toward their resolution;

• 71 percent said they personally were either active in or sympathetic to the environmental movement (notably, only 5 percent were "unsympathetic");

• In the same poll, 55 percent felt that environmental problems overall were either "very" serious or "extremely" serious (just 5 percent said "not serious at all"). Concern for nearly a dozen different environmental issues, including air and water pollution, ranked high, as well, with majorities reporting that they worried "a great deal" about most;

• Finally, despite thirty years of environmental legislation and regulation, 58 percent told Gallup that the U.S. government was currently doing "too little" to resolve those problems (merely 10 percent said it was doing "too much");[15]

In general, the asymmetry of those results appears consistent with Kempton, Boster, and Hartley's claim that there is no "coherent and consistent antienvironmental position" in American politics today.[16] Survey respondents do disagree at times—as do unemployed sawmill workers, dry cleaners, and environmental activists—but dissension, they say, generally arises over the "relative ranking" of environmental goals in comparison to other social values and priorities "rather than disagreement on the values themselves."[17]

It is a distinction with which Richard Ellis and Fred Thompson initially agree. In writing their own study of culture and the environment in the Pacific Northwest, they confirm that "it is not that activists want environmental amenities that the general public does not want—quite the contrary. Both want cleaner air and water, wilderness and species preservation, habitat protection, and a healthier, safer earth." Yet given that the essence of politics is often defined by the need to make difficult choices, they add that it is equally important to acknowledge that groups often disagree over the means used to achieve those goals.[18]

In short, how far does cultural consensus on the environment extend? Shared agreement, after all, is easier to achieve when survey questions "do not require choices between conflicting values," something Kempton, Boster and Hartley do admit.[19] Yet several of the measures they developed seem almost destined to generate agreement, including:

7. People have a right to clean air and clean water....

27. We have a moral duty to leave the earth in as good or better shape than we found it....

109. Nature may be resilient, but it can absorb only so much damage....

111. Working to try to prevent environmental damage for the future is really part of being a good parent.

In failing to press respondents further in their beliefs, Bron Taylor wondered whether Kempton and his colleagues had perhaps "made too much of respondent agreement with their survey propositions," perhaps using their results to show "little more than a recitation of empty truisms that bear little relation to environmental action."[20]

An extensive battery of questions included in the National Opinion Research Center's (NORC) General Social Survey (GSS) in 1994 suggests that Taylor's criticisms might be pointed in the right direction. Americans are genuinely concerned about the environment, and most appear willing to accept some form of government regulation and intervention in business and private decisions to aid in its protection. But as attention shifts from the general to the specific, consensus stops there. When confronted directly with the cost of protecting the environment—either through higher prices, increased taxes, or a reduction in the standard of living—respondents are plainly divided (see table 1.1).

Among the three considerations outlined in table 1.1—higher prices, higher taxes, and cuts in the standard of living, the issue of taxation is perhaps the most telling. A majority of those polled thought that "too

Table 1.1
Attitudes toward Environmental Costs and Trade-offs

Question	Agree	Neutral	Disagree
We worry too much about the future of the environment and not enough about prices and jobs today.	42%	14%	44%
People worry too much about human progress harming the environment.	37	15	48
To protect the environment, America needs economic growth.	45	26	44

Question	Willing	Neutral	Unwilling
How willing would you be to pay much higher prices to protect the environment?	47	24	28
And how willing would you be to pay much higher taxes to protect the environment?	34	21	44
And how willing would you be to accept cuts in your standard of living to protect the environment?	32	23	45

Source: NORC General Social Survey (1994).

little" was being spent by the government to protect the environment (61 percent), and yet only a third were willing to pay "much higher taxes" to fund those efforts. Pushed still further, respondents were divided, too, about the priority that environmental issues should be given relative to economic progress and prosperity—for instance, "We worry too much about the future of the environment and not enough about prices and jobs today." Cultural consensus, it seems, does not always lead to political consensus.

Setting Priorities, Balancing Goals

While differentiating the means of public policy from its ends clearly matters when polling Americans on the environment, the disagreement that Kempton and his colleagues find over the ranking of environmental

Table 1.2
Ranking the Priority of Environmental Protection

First, I'd like to read you some issues the country will be facing over the next few years. For each one, please tell me how much of a priority it is for the country to address—using a scale of 1 to 10—where 10 means something should have high priority and 1 means it should have the lowest priority. You may use any number from 1 to 10. First how high a priority should _____ have? [ROTATE.]

Rank	Question Item	Mean Response on 10-Point Scale	Percentage Responding 10 (high)
1	Lowering crime rates	8.80	58%
2	Improving public education	8.73	55
3	Improving the economy	8.37	42
4	Improving the health care system	8.24	43
5	Cutting government spending	8.16	46
6	Protecting the environment	7.97	38

Source: Belden and Russonello (1996).

issues relative to other things raises a second caveat as well—whether it is more important in survey research to weigh the nominal support an issue receives or the rank order into which it is placed. A short example illustrates the difference.

In the winter of 1996, a consortium of research foundations called the Consultative Group on Biological Diversity (CGBD) sponsored a public opinion poll to gather information on American attitudes toward loss of species and their habitats, as well as toward the environment in general.[21] At first glance, the results of that survey were clearly impressive. On a scale ranging from one to ten, the mean priority that respondents placed on "protecting the environment" was a definitive eight, and when asked about maintaining biological diversity in particular, an overwhelming 87 percent of those polled felt it was an important goal.[22]

Both conclusions, however, are less convincing when returned to their original context. In this case, the questionnaire developed for the CGBD asked respondents to rate the priority of environment protection, as well as other issues, including crime, health care, public education, and the economy. As table 1.2 points out, when the environment is compared to

Table 1.3

Ranking Public Concern for Biodiversity

Thinking specifically about environmental issues, please tell me how serious a problem you think each of the following is, using a scale of 1 to 10 where 1 means something is not a problem at all and 10 means it is a very serious problem.

Question	Mean Response on 10-Point Scale	Percentage Responding 10
Toxic waste in the United States	7.85	36%
Loss of rain forests	7.70	35
The rate at which land is being developed and places in nature are being lost	7.66	33
Rate of growth of the world's population	7.41	29
Air quality in the United States	7.24	24
Water quality in the United States	7.24	25
Overconsumption of resources in the United States	7.24	23
The rate at which plant and animal species are becoming extinct	6.89	25
Acid rain in the United States	6.36	14
Global climate change	6.20	15

Source: Belden and Russonello (1996).

those topics, its average score on a ten-point scale is far less imposing, actually ranking last on the list in descending order. This result has been replicated time and again by other polls under similar conditions.[23]

Likewise, concern for the seriousness of species loss seems solid—a credible 6.9 on a ten-point scale—until that issue also is placed beside other environmental problems, including air and water pollution, population growth, and toxic waste disposal (see table 1.3). In the end, a fair interpretation of the poll would need to note that plant and animal extinctions ranked near the bottom of a list of environmental problems, which themselves rank at the bottom of a broader list of social priorities. Americans are concerned about the environment, to be sure, but they often find other issues more pressing and immediate. In politics, of course, it is the latter that forces action.

Loading the Dice

Finally, while survey results like those cited above can be taken out of context by the news media and can unintentionally mislead readers, some surveys are deliberately designed to do so.[24] Consider the experience of Rolla Williams, an outdoor staff writer for the *San Diego Union-Tribune*, who in 1987 received a questionnaire in the mail from the Sierra Club. Among the questions posed in that survey were these:

Our nation is still blessed with millions of acres of public lands, including roadless wilderness areas, forests, and range lands. Land developers, loggers, and mining and oil companies want to increase their operations on these public lands. Do you think these remaining pristine areas of your public lands should be protected from such exploitation?

Sulphur dioxide from industrial smokestacks has caused the phenomenon called acid rain. This rapidly growing threat to our forests, lakes, and human health has already killed thousands of lakes in the United States and Canada. The administration in Washington has argued that we don't know enough about the problem to devise an appropriate remedy. The Sierra Club views acid rain as one of the most serious environmental problems facing us today. What priority would you give to attacking this problem?

As Williams notes, "How does one answer" such questions? Given the survey's failure to reflect the complex realities of environmental policy, he wryly suggests that that the Sierra Club "hire the Gallup Organization to draw up its next questionnaire and remove every vestige of the bias" found within it.[25]

Scholars, of course, have long recognized that the way questions are asked in surveys matters.[26] Unbalanced assertions, double-barreled questions, and leading phrases all have the potential to introduce response sensitivity. Identifying those problems (and their complementary solutions) is easier, perhaps, through comparison. In a 1989 cross-national study of environmental attitudes administered by Louis Harris and Associates, respondents in sixteen countries were asked a lengthy number of questions that included the following:[27]

4. How concerned are you that _____? Are you very concerned, somewhat concerned, not very concerned, or not at all concerned? [ROTATE.]
1) The air you breathe is becoming less healthy;
2) The water you drink is becoming less safe and a danger to health;
3) Lakes and rivers are being polluted by man-made chemicals from industry;

5) The chemicals used to control pests and weeds are making food and water supplies unsafe;
9) Radiation from nuclear power plants will escape and kill thousands of people;
11) Dangerous chemicals are being dumped by industry without taking safety precautions to protect people from being poisoned;

Text provided in table 1.4 pairs the Harris questions as closely as possible to those used in a standard Gallup battery that same year. The differences in language and syntax are instantly apparent.

In using active verbs that include both present and future tenses, the questions developed by Harris do not use a neutral tone to ask respondents to assess their concerns but rather to respond to problems that are presumed to be both real and threatening. For instance, while Gallup inquires about personal worry about the "contamination of soil and water by toxic waste," Harris asks respondents how concerned they are that "dangerous chemicals are being dumped by industry without taking safety precautions to protect people from being poisoned." Under those conditions it is not surprising that majorities in all countries responded to the Harris questions by saying that they were not only concerned but "very concerned" about those particular issues.

Despite a clear bias, data results from the Harris study were widely disseminated to the news media. In commenting about the poll's release at the time, Mostafa Tolba, executive director of the United Nations Environment Programme (UNEP) said, "We are very encouraged to see the strength and the depth of support for both national and multinational environment programmes. We have a clear mandate for our work. I hope the survey will be seen as a call to action."[28]

Politics, the Polls, and Global Warming

Up to this point, three warnings about measuring the direction of environmental concern in public opinion polls have been noted:

• The importance of distinguishing support for policy ends from policy means;

• The advantage of weighing the implied rank of social priorities in addition to their nominal value; and

• The need to consider potential biases in question wording and format.

Table 1.4
A Comparison of Question Wording

Gallup Organization	Louis Harris and Associates
[P]lease tell me if you personally worry about this problem a great deal, a fair amount, only a little, or not at all.	*How concerned are you that _____? Are you very concerned, somewhat concerned, not very concerned, or not at all concerned?*
Pollution of drinking water	The water you drink is becoming less safe and a danger to health.
Pollution of rivers, lakes, and reservoirs	That lakes and rivers are being polluted by man-made chemicals from industry.
Contamination of soil and water by toxic waste	Dangerous chemicals are being dumped by industry without taking safety precautions to protect people from being poisoned.
Air pollution	The air you breathe is becoming less healthy.
Contamination of soil and water by radioactivity from nuclear facilities	Radiation from nuclear power plants will escape and kill thousands of people.
The "greenhouse effect" or global warming	Chemicals from factories and cars are going into the atmosphere and making the climate worse.
Extinction of plant and animal species	Many types of animals, birds, fish, insects, and plants are dying off.
Acid rain	Chemicals from industries and factories are causing acid rain, which is killing forests and life in many lakes.

Source: Gallup Organization (April 3–9, 2000) (*n* = 1,004); Louis Harris and Associates, Inc. (1989).

When combined, how severely can these complications influence our understanding of public support for environmental protection? A return to the issue of global warming provides the answer.

In August 1997, several months before the results of the Pew Research Center poll reached the news media, the World Wildlife Fund (WWF) sponsored a survey of its own on the "greenhouse effect." Interviewing eight hundred registered voters nationwide, the press release accompanying the WWF poll results stated that the poll had revealed "great enthusiasm for international efforts to address global warming." Rejecting the belief that reducing carbon emissions would disrupt the economy and cost jobs, respondents demanded instead that political leaders "act immediately" on the issue.[29]

While these results at first glance appear consistent with the Pew study and with a host of other similar surveys conducted in the months leading up to the Kyoto Protocol in Japan, a number of problems come into play.

Policy Discord

Most polls conducted during the summer and fall of 1997 found high levels of public concern for the issue of global warming, and the WWF poll was no exception to that rule. Asked "how serious a threat" they thought global warming was at that time, 45 percent of those responding believed it was "somewhat" serious, while an additional 29 percent categorized it as "very serious." Only 7 percent felt it was "not serious at all."[30] In response, the World Wildlife Fund concluded: "Despite the conventional wisdom that only radical environmentalists and science fiction fanatics are concerned about global warming, our recent survey reveals that the American public believes global warming is real and represents a serious threat."[31]

That community of opinion, however, did not extend to the policies proposed to address the problem. Available data suggest two reasons why. First, while public concern is significant, there can be genuine ideological differences of opinion on issues of public policy, which in the environmental field often revolve around the degree of government intervention in the market economy. For example, the WWF survey asked,

In trying to reduce the threat of global warming, do you think we should rely mainly on strict regulations to limit emissions of carbon dioxide, or do you think we should rely mainly on incentives that will cause the free market to discourage carbon dioxide pollution—or don't you have an opinion on this?

In response to that question, 37 percent of respondents favored government regulations, while 32 percent preferred free-market options and incentives. Yet the high proportion of "don't know" responses to the question (the remaining 30 percent) suggests a second explanation as well—that the WWF survey contributed to response instability by asking average citizens to evaluate scientific conditions and complex policy proposals about which they had little knowledge, experience, or information. It is a point well reinforced by the battery of questions outlined in table 1.5, where lay respondents were asked to judge "how likely" specific problems were to occur as a result of global warming trends.[32]

In truth, most Americans know little about climate change.[33] The Pew Research Center survey wisely asked respondents, based on what they had heard or read (if anything), how they would describe the "greenhouse effect." More than a third of those polled (38 percent) could not define the concept even in the vaguest of terms, identifying it instead, when presented with a close-ended list of options, as either a "new advance in agriculture" or a "new architectural style" rather than an "environmental danger."[34] A similar result occurred in the NORC 1994 General Social Survey (GSS), where more than half of those polled (55 percent) believed, incorrectly, that the greenhouse effect was caused by a hole in the earth's atmosphere.[35]

Those results are more understandable, perhaps, when compared to the attention people paid to a variety of news stories that year. Solid majorities in the Pew study reported that they had followed "Iraq's refusal to let Americans participate in weapons inspections" (76 percent) closely, the Massachusetts murder trial of British au pair Louise Woodward (65 percent), the ups and downs of the stock market (61 percent), and even the flooding produced by the weather phenomenon, El Niño (62 percent). In contrast, just a third said the same about the debate over global warming (ranking lowest on a list of eleven news topics), despite a renewed concentration of media attention in the months leading up to the Kyoto Protocol.[36]

Table 1.5
Public Opinion on Global Warming

I'm going to list some specific problems that some people say could happen as a result of global warming. After each, please tell me how likely you think that problem is to occur as a result of global warming. Is that problem almost certain to happen [4], very likely [3], somewhat likely [2], not too likely [1], or not likely at all [0] to occur as a result of global warming? If you are not sure how you feel about a particular item, please say so, and we will go on.

Question	Mean likelihood of happening
More extreme weather conditions, such as drought, blizzards, and hurricanes	2.41
Longer and hotter heat waves leading to more heat-related deaths	2.36
Crop failures, food shortages, and famine caused by droughts and floods	2.34
Melting ice caps and glaciers causing the sea level to rise and the flooding of coastal communities	2.34
Disruption of our fragile ecological balance threatening the web of life	2.25
Destruction of natural habitats, including oceans and national parks like Glacier National Park and the Everglades	2.22
Increased incidences of asthma and other respiratory problems	2.19
Extinction of certain animal species like polar bears and sea turtles	2.19
The spread of infectious diseases	1.76
Increased populations of pests and vermin, such as roaches, termites, and rats	1.74

Source: World Wildlife Fund (1997a).
Note: The above battery was asked only of respondents in split sample B. $n = 800$.

With that deficit of attention and information in mind, the complexity of the issues raised in the WWF survey is particularly striking. For example, respondents in the poll were asked to consider the following question:

I'm going to list some of the specific proposals that have been made to decrease the use of oil, coal, and gasoline in order to reduce the threat of global warming. For each item I read, please tell me if you favor or oppose each one. If you aren't sure how you feel about any specific item, just say so and we will go on. . . .
Have the United Nations establish a worldwide limit on carbon dioxide emissions that is lower than current levels. Each U.N. member country would be allocated the right to discharge a certain amount of carbon dioxide pollution. Countries could buy and sell these pollution rights to one another. This would allow them to choose between reducing their carbon dioxide emissions or paying to continue to pollute.

As Michael Traugott and Paul Lavrakas explain, many respondents find it difficult to answer complex, policy-driven questions that reach beyond their own knowledge and life experience. The responses to measures such as these, they say, are not very informative in the end because they might represent little more than "a respondent's pure guess about what the question means."[37]

A broader comparison of questions used by various polling organizations in 1997 illustrates a similar point. In each case, while an American decision about how to respond in Kyoto hung in the balance, respondents were asked a variant of the same basic question: should the United States take immediate action on global warming or not? The results were decidedly inconsistent.

In the World Wildlife Fund survey, 56 percent of those polled urged President Clinton to "take action on global warming now," but an equal number in an NBC News/Wall Street Journal study believed that "more research is necessary before we take action" (59 percent). In a CBS News/New York Times poll, 81 percent said steps to counter the effects of global warming should be taken "right away," but Charlton Research Company (1997) counted 78 percent who believed that the United States should "wait to make any treaty commitments" and pursue only "voluntary programs" instead.[38] Underscored by a lack of knowledge and information, the nature of the options given to participants in each of these cases made all the difference in the world.

Mistaken Priorities

Even if Americans were understandably confused and divided about how best to approach the issue of global warming, was not their concern genuine? In many ways, it was. Time and again polls have shown that a majority of Americans believed that the "greenhouse effect" was real and that its consequences were serious. Yet given the natural tendency of Americans to express concern for issues of all kinds when asked by eager pollsters, those results require some context—a baseline against which to compare and contrast global warming with other issues. Several studies (including Pew's 1997 survey) provided just that, showing that global warming ranked comparatively low for most respondents, in fact well below air and water pollution, habitat loss for wildlife, and destruction of the rain forests. Still, its implied priority on that list ultimately did little to redirect media attention away from the sensational. Although James Gerstenzang, a columnist for the *Los Angeles Times*, admitted in print that the data showed "most people do not consider the climate debate urgent," the headline of his article struck a decidedly different tone, claiming "Survey Bolsters Global Warming Fight."[39]

Leading Questions

Finally, on issues such as global warming, where opinions tend to be ill informed, low key, and unstable, question wording can become a crucial determinant of the results. In the case of the WWF survey, certain questions were not just confusing; they were biased and misleading. For instance, respondents were asked which of the following two viewpoints they agreed with more:

Some people say that most scientists agree that global warming is real and already happening. They say the only scientists who do not believe global warming is happening are paid by big oil, coal, and gasoline companies to find the results that will protect business interests, just like the tobacco industry scientists who said cigarettes don't cause cancer. [75 percent]

Other people say that scientists disagree among themselves that global warming is happening. They say there is no real evidence that carbon dioxide emissions from coal, oil, and gasoline are causing global warming. [15 percent]

Notice how framing the debate over global warming by reference to another well-known controversy—in this case, one involving the tobacco industry—compromises the tone and content of the question. Here, respondents were encouraged to equate two issues that may (or may not) share an objective similarity. As Karlyn Bowman points out, aside from potentially skewing the results, questions like those above also "undermine the credibility of the polling profession" and lend credence to the unfortunate belief that survey research does more to shape public opinion than to measure it.[40]

Solid Ground

Given all of these difficulties, where do Americans stand on the issue of global warming? As Bowman argues, the answer is to be found in the consistency of responses across multiple questions administered by different polling organizations over time. Those areas of agreement suggest that a majority of people believe that global warming is a real phenomenon with serious environmental consequences, but the concern they feel fails to reach a level of alarm, which may help to explain why citizens are so undecided about the pace of government efforts to reduce emissions of greenhouse gases. John Immerwahr offers an additional insight as well. On the issue of global climate change, he says, Americans "run into brick walls, characterized by lack of clear knowledge, seemingly irreversible causes, and a problem with no real solution." As a result, respondents are "eager for a solution but unsure of which way to go," often preferring to wait and see before committing U.S. resources and taxpayer dollars to a painful, corrective course of action.[41]

Conclusions

To observe the direction of public opinion is to start at the very beginning, to reduce the study of environmental attitudes down to its barest essentials.[42] At first glance, it might appear to be little more than the challenge of delineating a rough dichotomy—either Americans are concerned about the environment, or they are not. But as this chapter has argued, the appearance of consensus in surveys tends to coexist alongside expressions of genuine ambivalence. It is an uneasy combination that paints a complicated picture.

For example, as the debate over global warming shows, Americans care about the environment but frequently send mixed signals to policymakers by placing priority on other issues. Moreover, shared agreement on environmental values and concerns often fails to extend to the means used to achieve those goals. In some cases, taxpayers may be reluctant to accept fiscal responsibility for their demands; in others they seem to cleave over larger ideological debates surrounding governments and markets. Americans, too, know little about the intricacies of public policy, and so when they are asked to evaluate political choices, their responses can be unstable and contingent on the manner in which questions are phrased in surveys. In a way, all of these factors help to explain lingering perceptions of environmental consensus. As long as dialogue focuses on broad goals, diverse populations can be unified by a similarity of experience without necessarily agreeing on what steps should be taken to ensure environmental quality.

In the end, however, perception alone is insufficient, even in politics. Each of these findings underscores a certain poverty of language, demanding careful analysis. While the natural tendency of the news media is to simplify and condense, a firm understanding of public opinion is defined by far more than a single characteristic. Attitudes may be clustered or divided in their direction, but so too are they sorted and shaped on a mass scale by other properties, enumerated by V. O. Key to include (among others) strength, stability, distribution, and constraint—all of which provide the building blocks of the chapters to come.[43]

2

Strength: How Deep Is Public Commitment to the Environment?

In 1900, while writing the second volume of his classic book *The American Commonwealth*, James Bryce insisted that the term *public opinion* referred to something greater than the sum of its parts. He saw it as an interactive consensus, the endpoint of a lengthy process of social and political debate. For Bryce, to count only the number of citizens who aligned themselves directionally on either side of an issue (weighing each of those opinions equally) meant missing an important part of the picture. Even in an era long before the advent of modern survey research, scholars recognized that there was an important distinction to be made. As A. Lawrence Lowell argued in 1913, "One man who holds his belief tenaciously counts for as much as several men who hold theirs weakly," for as he reasoned, "individual views are always to some extent weighed as well as counted."[1]

Today, our ability to weigh opinions using the science of social research is greater than ever before, but the political consequences of that simple insight still resonate. V. O. Key, for instance, noted that "as influences work themselves out through political processes, the qualities of an opinion, as well as the numbers who hold it, enter into the equation," including first and foremost the notion of attitude strength. "It requires no monstrous stretch of the imagination," he said, "to suppose that this fact seeps through the mechanisms of communication and in one way or another affects the course of political action."[2] Key believed that under some conditions passionate attitudes had the ability to magnify small numbers, while at other times lukewarm commitment could dampen the significance of widespread support.

For the contemporary environmental movement, of course, the parallel is unmistakable and rather vexing. While most Americans are broadly sympathetic to environmental goals, scholars have long cautioned that those attitudes may be weakly held.[3] As a result, environmentalists looking to capitalize on grassroots support politically are faced with the constant challenge of energizing their base. They must, after all, convince average citizens, already overwhelmed by the demands of daily life, that environmental problems are worthy of scarce energy and attention and that they should not just care about the environment but care deeply. In that vein, gauging attitude strength is important to far more than the study of public opinion alone. It becomes a necessary ingredient in a viable political strategy.

With those stakes in mind, differentiating between numbers and intensity will be the foremost challenge of this chapter. While it is one that highlights (yet again) the difficulty scholars have in polling Americans on the environment—not only on the subject of what they think, but also how they behave—one basic point survives a multitude of complaints about question wording and survey design. Despite Americans' reservoir of good will and honest concern, when pressed, their commitment to environmental protection is equivocal when and where it matters most.

Measuring Attitude Strength

The development of valid and reliable measures of attitude strength has generated a great deal of experimentation in the field of survey research, and yet it is a concept that remains problematic in practice, especially on issues like the environment, where social desirability biases loom large.[4] In *Questions and Answers in Attitude Surveys* (1981), Howard Schuman and Stanley Presser explore the problem of attitude strength by reference to two subjective qualities in particular—intensity and centrality.[5] *Intensity*, they note, refers to a "strength of feeling," while *centrality* is defined by a sense of "importance." A third category can be added to the mix as well, something Schuman and Presser label *conviction* of opinion, where counterarguments and paired comparisons (or trade-offs) are introduced. Each of these factors, individually and in combination, is explored below using a number of recent environmental polls.

Intensity of Opinion

The most direct way to measure attitude intensity (or strength of feeling) is perhaps through the use of standard degree-of-concern items that are capable of sifting black and white responses to complex problems into several different shades of gray. It is a common technique that has been used by the Gallup Organization in its environmental battery since 1989. In those questionnaires, respondents are asked if they personally worry "a great deal," "a fair amount," "only a little," or "not at all" about each of a number of environmental problems. The distribution of those responses across categories, as found in a March 2001 version of that survey, is shown in table 2.1. The results tallied there suggest that Americans do harbor serious anxiety about the environment. In fact, on many of the issues described by Gallup, a majority of those polled gravitated toward the high end of the scale.

As straightforward as this approach may seem, however, it is not without critics. Arthur Sterngold, Rex Warland, and Robert Herrmann argue that degree-of-concern items often contain unstated assumptions. By presuming that people are (or should be) concerned about the subject at hand, respondents can be subtly led by unbalanced questions to "accommodate this expectation by overstating their actual concerns." In a split-ballot experiment testing their hypothesis, Sterngold and his colleagues found that groups receiving an initial filter question gave roughly double the percentage of "not concerned" responses, while those with genuine concern were less likely to place themselves at the upper end of the response scale.[6] When these issues of question format are paired with the natural tendency of environmental issues to fall prey to social desirability biases, Everett Carll Ladd and Karlyn Bowman express skepticism about expressions of concern that come free of consequence. "Viewed in isolation," they say, "the results to these questions seem to suggest enormous concern. In fact they merely show that many people do not like pollution."[7]

While Ladd and Bowman's criticism may be accurate, they overlook other important inferences on attitude strength that can be found in lengthy batteries like those used by Gallup. For example, biases that press respondents to inflate their true sense of environmental concern may be worrisome, but they should leave comparisons *between* issues

Table 2.1
Concern for Thirteen Environmental Problems

I'm going to read you a list of environmental problems. As I read each one, please tell me if you personally worry about this problem a great deal [4], a fair amount [3], only a little [2], or not at all [1]. First, how much do you personally worry about? [RANDOM ORDER]

Question	Not at all	Only a little	A fair amount	A great deal
Pollution of drinking water	2%	9%	26%	62%
Pollution of rivers, lakes, and reservoirs	2	11	29	58
Contamination of soil and water by toxic waste	3	13	28	56
Air pollution	4	15	34	47
Contamination of soil and water by radioactivity from nuclear facilities	11	20	22	47
Urban sprawl and loss of open spaces	9	20	35	46
The loss of natural habitat for wildlife	4	17	33	46
The loss of tropical rain forests	8	16	31	45
Damage to the earth's ozone layer	9	17	29	45
Ocean and beach pollution	4	17	35	43
Extinction of plant and animal species	8	20	31	42
The "greenhouse effect" or global warming	13	22	31	34
Acid rain	17	27	29	28

Source: Gallup Organization (March 5–7, 2001).
Notes: $n = 1,060$ adults nationwide. Margin of error ± 3 percentage points.

virtually unaffected, and as chapter 1 notes, comparison is a useful tool indeed.[8] For example, on several issues included in the Gallup study, respondents shifted collectively toward the moderate range of the scale, most notably on threats that are (at least relative to pollution) more controversial within the scientific community and less visible to the untrained eye, including ozone depletion, global warming, and acid rain. It may be difficult in the end to judge precisely *how* concerned Americans are about the environment by reference to the environment alone, but it seems reasonable to use those measures to speak in terms of "more" or "less."

A related (but less hazardous) technique for measuring attitude intensity is the traditional Likert scale, where participants are asked to react to a given statement by attaching themselves to a well-balanced response format that includes labels such as "strongly agree," "somewhat agree," "somewhat disagree," and "strongly disagree," often accompanied by a neutral position that serves as midpoint.[9] A good illustration of that technique is a question asked below of respondents (both with and without the Likert format) by Wirthlin Worldwide and CBS News/New York Times, respectively:[10]

[Please tell me whether you agree or disagree with the following statement:] Protecting the environment is so important that requirements and standards cannot be too high, and continuing environmental improvements must be made regardless of cost.

Wirthlin Worldwide (1998)

Strongly agree	31%
Somewhat agree	32
Somewhat disagree	21
Strongly disagree	15
Don't know	—

CBS/New York Times (1997)

Agree	57%
Disagree	36
Don't know	7

Note how dramatically different the results of these two surveys seem, and yet they are entirely consistent. Roughly the same proportion of respondents align themselves directionally on either side of the issue, and yet the addition of a Likert-type response scale in the Wirthlin poll introduces a sense of weakened resolve and genuine ambivalence. At length, questions of this kind are preferable to directional formats and standard degree-of-concern items because they are more demanding and, as such, less susceptible to well-meaning exaggeration.

Centrality of Opinion

While a more pragmatic picture begins to emerge in the Wirthlin study, it is important to recognize that the placement of an opinion along an ordinal response scale is not a sole indicator of attitude strength. As Schuman and Presser write, measures of intensity often fail to represent the level of subjective importance that individuals place on the issue at hand.[11] They argue that the "centrality" of attitudes should be considered independently. Given all of the potential biases outlined earlier, one popular way to do that is to rely on unprompted, open-ended answers to questions that ask respondents to name the nation's "most important problem." Under those conditions, the environment fares badly indeed: it is mentioned by just 2 percent of those polled in an April 2000 Gallup poll—a figure not appreciably higher or lower than that found in most other surveys over the past thirty years.

As Riley Dunlap argues, however, "most important problem" questions can be problematic indictors of attitude strength.[12] Some scholars find that responses are particularly susceptible to news coverage. Given that media attention to environmental issues is not generally high (disasters like the *Exxon Valdez* oil spill notwithstanding), trends in those questions may simply reflect the influence of media agenda setting.[13] In addition, as Robert Cameron Mitchell writes, "Mass salience is transitory for all but the most momentous issues such as war or depression," making the format unsuitable for less dramatic events.[14] In short, if degree-of-concern items make it too easy for respondents to express worry about almost anything, "most important problem" questions seem to err on the side of stringency.

In response, Dunlap once again promotes the use of comparison, citing close-ended response formats as a "more sensible" solution.[15] As table 2.2 shows, survey results in this area are more favorable to the environment, but regardless of the design, environmental issues "still fail to reach the very top of the list."[16] Quite simply, in weighing an accumulation of evidence, Ladd and Bowman come to the conclusion that "other problems are far more urgent than the environment for Americans today."[17]

Conviction of Opinion

While measures of intensity and centrality can be important indicators of attitude strength (or weakness), the inherent social desirability of environmental goals encourages the use of a third assortment of questions as well, commonly referred to as paired comparisons or "trade-offs."[18]

Table 2.2
Perceived Seriousness of Various Social Issues

Next, I am going to read a list of problems currently facing our country. For each one, please tell me how serious of a problem you consider it to be for our country—extremely serious [4], very serious [3], somewhat serious [2], or not serious at all [1]. [0 = no opinion] [RANDOM ORDER].

Rank	Question item	Mean response on item scale	Percentage responding "extremely serious"
1	Drug use	3.19	38%
2	Crime and violence	3.14	33
3	Poor health care	2.91	29
4	Hunger and homelessness	2.86	24
5	Environmental problems	2.65	17
6	Racial conflict	2.62	18
7	Illegal immigration	2.48	15

Source: Gallup Organization (April 3–9, 2000).
Notes: n = 1,004 adults nationwide. Margin of error ±3 percentage points.

Consistent with Schuman and Presser's definition of *conviction* of opinion, this approach is designed to examine the ease at which citizens are influenced by counterarguments and opposing goals, all of which force them to consider the very real costs associated with protective environmental policies.[19] For example, in nearly every year since 1984 the Gallup Organization has asked respondents the following question:[20]

With which one of these statements about the environment and the economy do you most agree?
A. Protection of the environment should be given priority, even at the risk of curbing economic growth; [67 percent]
B. Economic growth should be given priority, even if the environment suffers to some extent. [28 percent]

This is an improvement, perhaps, considering Ladd and Bowman's initial criticism regarding degree-of-concern items, and yet trade-off questions also earn their disapproval. As they explain, because most people "firmly believe that we can have both a clean environment and economic growth, questions that ask Americans to choose between one and the other are highly misleading."[21]

To demonstrate their point, consider the following pair of questions asked of split half-samples in a 1994 Cambridge Research Reports International poll:[22]

Which of these two statements comes closer to your opinion?
We must be prepared to sacrifice environmental quality for economic growth, or we must sacrifice economic growth in order to preserve and protect the environment."

Sacrifice economic growth   53%
Sacrifice environmental quality   23
Don't know   24

Which of these two statements is closer to your opinion?
There is no relationship between economic growth and the quality of the environment. Indeed, we can have more and more goods and services and also a cleaner world, or we cannot have both economic growth and a high level of environmental quality. We must sacrifice one or the other.

Can have both   67%
Cannot have both   25
Don't know   8

Given that each measure offers a very different answer to the same underlying issue, Ladd argues that trade-off questions often present respondents with forced choices that offer "nothing more than differing pieces of one basic value that the public wants, and believes it possible to attain."[23]

Still, Ladd's criticism overstates a limited point. Trade-off questions do force an unpleasant choice that most respondents (naturally) hope to avoid, but issues involving competing values are not always escapable. As Scott Keeter notes, politics is precisely about the kinds of choices required by trade-off questions, and "while the public is not well equipped to make elaborate calculations of the costs and benefits of particular environmental policies," those comparisons remain one of the most rigorous ways of determining "how the public ranks certain values."[24]

In short, answers to trade-off questions speak not simply to the authenticity of the options offered but to the conviction with which citizens hold their environmental beliefs. While Ladd and Bowman fear that confusion over the real intent of the question means "it is impossible to know what to make of the answers given," Schuman and Presser's extensive experiments on question wording and format suggest otherwise.[25] They write that one of the "clearest findings" of their work "is the extent to which people, once they have agreed to be interviewed, accept the framework of questions" that they are offered and "try earnestly" to work within it.[26]

More and Better through Combination

While measuring attitude strength can be difficult, as each of the above examples testifies, it is not an impossible task, especially when the most promising elements of intensity, centrality, and conviction are combined deliberately and skillfully into a single survey. This was surely the case in the 1996 National Election Study (NES). Using a seven-point numerical scale, respondents in the pre-election poll were asked to place themselves on a continuum anchored on each end by desirable goals described as "protecting the environment" and "maintaining jobs and our standard of living." This formulation minimizes several of the dangers noted above.

First and foremost, in contrast to standard degree-of-concern items, the advantage of environmental protection in terms of social desirability is to some degree neutralized in this measure through direct comparison to another advantageous goal, economic prosperity. Moreover, both goals were worded in a reasonable way, one that average Americans were likely to comprehend without reference to complex policy options or debates.

Second, while trade-off questions can be criticized for presenting false choices, the odd-numbered scale allows an implied midpoint representing a position of neutrality and moderation in choosing between two goals that respondents may very well have believed could be achieved in tandem.

Finally, two questions asked immediately following the initial scale offer additional insights. The first asked respondents how certain they were of their position on the scale, and the second, using a Likert-type response format, asked how "important" the issue was to them. In this context, the former question speaks to attitude crystallization, while the latter becomes an effective indicator of centrality and yet one not as unfairly demanding as the typical "most important problem" question.

Table 2.3
Exploring Attitude Strength on the Environment

[Issue position:] *Some people think it is important to protect the environment even if it costs some jobs or otherwise reduces our standard of living. (Suppose these people are at one end of the scale, at point number 1.) Other people think that protecting the environment is not as important as maintaining jobs and our standard of living. (Suppose these people are at the other end of the scale, at point number 7.) And, of course, some other people have opinions somewhere in between, at points 2, 3, 4, 5, or 6. Where would you place yourself on this scale, or haven't you thought much about this?* [V960523]

[Issue certainty:] *How certain are you of your opinion on this scale: (1) not very certain, (2) pretty certain, or (3) very certain?* [V960524]

[Issue importance:] *How important is this issue to you: (5) extremely important, (4) very important, (3) somewhat important, (2) not too important, or (1) not important at all?* [V960525]

Rank	Issue Position	Aligned on scale	Mean certainty	Mean importance
1	Protect environment, even if it costs jobs, standard of living	11.6%	2.78	4.46
2		15.7	2.51	4.13
3		20.2	2.31	3.74
4		27.8	2.17	3.51
5		14.0	2.25	3.47
6		6.9	2.49	3.78
7	Jobs, standard of living more important than environment	3.9	2.77	4.02

Source: National Election Study (1996).

Although not startling, one of the most meaningful conclusions to be drawn about attitude strength on the environment can be found in the combination of these three items. While in most degree-of-concern questions respondents cluster toward the extreme, nearly two-thirds of those polled by NES gravitated to the middle range of the issue scale—outlined in table 2.3—creating a distribution that has all the appearance of a bell curve.

As might be expected, respondents falling to the center of the scale were also somewhat less sure of their position and found the issue to be of less importance to them personally.[27] In short, all three questions reinforce the same basic point: Americans may be concerned about the environment, but many are ambivalent and unresolved about how strongly those issues should be addressed when other values enter into the equation.

From Words to Action

The difficulties that researchers encounter in finding suitable measures of attitude strength suggest one glaring omission from the list above. As the popular adage reminds us, "talk is cheap," and so those who feel strongly about something are often urged to put "their money where their mouth is."[28] The same might be said of the public's commitment to environmental protection. Why not, then, look beyond mere words to the realm of action and judge the depth of concern by our collective willingness to act on our stated beliefs as voters, consumers, or political activists? Unfortunately, this approach is no less difficult (or controversial) to navigate.

Contingent Valuation

Valuing the environment by reference to behavior is a difficult task for one essential reason: the objects at hand include collective, rather than private, goals. While certain environmental goods can be priced in indirect ways by measuring property values or travel costs to recreational areas, most environmental resources (such as clear air or water) are seldom, if ever, exchangeable in the open marketplace.[29] Economists caution, too, that such studies are likely to underestimate environmental preferences by accounting for use alone, dismissing as intangible other option or existence values.[30]

Recent trends in environmental economics, however, have wrestled with the measurement of intangible factors through the use of contingent valuation studies. Rather than attempting to observe actual market behavior using monetary measures, contingent valuation extracts personal preferences through polling by creating hypothetical conditions under which respondents state their "willingness-to-pay" (WTP) for an infinite variety of environmental goods, ranging from park land to sea birds.[31]

But, with that said, of course, the attempt to move from words into action ultimately returns us to square one—that is, to the problems and constraints of survey research.

Not surprisingly, results from these efforts have been decidedly mixed. While the NORC General Social Survey (GSS) found that strong majorities supported increased spending for environmental protection, those numbers grew soft when attention turned to specifics, including reference to a higher tax burden. Although 72 percent of those polled in a 1993 Louis Harris poll expressed willingness to "pay somewhat higher taxes" provided "the money would be spent to protect the environment and prevent air and water pollution," just 34 percent were equally willing in a similar GSS question asked a short time later.[32] The key difference in the that case was probably the latter's substitution of the phrase "much higher taxes," even though both questions were notably ambiguous as to cost.

In a similar comparison, Cambridge Reports Research International found in 1994 that 40 percent of respondents were willing to some degree to pay 5 percent more in higher taxes in order to "better protect the environment." Just two years before, however, the polling firm of Yankelovich, Clancy, Shulman found a much stronger 64 percent willing to pay $200 for the same purpose.[33] Even though the dollar amount required in each scenario was similar for a typical American family, the contrast in results was likely due to the uncertainty in which cost was expressed in the Cambridge study. Quite simply, respondents might have judged a 5 percent increase in taxes to be either high or low, but the figure probably had little concrete meaning to them.

Finally, two polls conducted on the topic of global warming in 1997 show how inconsistent responses can be on willingness-to-pay higher gasoline prices. While Princeton Survey Associates found respondents split, with 51 percent willing to pay twelve cents more, a Pew Research Center poll found an even stronger proportion (60 percent) willing to accept an increase of twenty-five cents.[34]

As each of these examples suggests, the concept of contingent valuation is in itself fraught with potential problems. Respondents find it difficult to react to hypothetical situations. Without any real-world frame of reference, they are unlikely to know the true cost of incremental improvements in environmental quality and therefore may be unable to choose a level of monetary support that is compatible with their preferences.[35] As others argue, too, "subtle changes in question order and wording can affect the nature of the responses,"[36] leading at times to exaggerated "green generosity," or alternatively, to "free riding" when the contributions of others are anticipated.[37] Finally, it is important to note that willingness to pay seldom represents an eagerness to do so. As Ladd reasons, responses to hypothetical questions regarding spending and taxation must not be understood "literally" but rather as an entirely "symbolic" commitment to environmental protection.[38]

Self-Reported Behavior

To gauge public commitment in ways that move beyond symbolism, a somewhat different approach is needed, one that refocuses attention not on support for collective public outcomes but rather on private, individual decisions. The latter point is addressed in an April 2000 Gallup poll (see table 2.4). In that questionnaire, respondents were asked which of a series of environmental tasks they had participated in "in the past year." While the results of that list are initially impressive, at least one note of warning is necessary. From a methodological standpoint, the validity and reliability of self-reported behavior can be at risk when respondents exaggerate or misrepresent their participation in certain activities.[39] As Stanley Presser writes, "People like to see themselves as good citizens or, more generally, to present themselves in a socially desirable light."[40] Given recent evidence that environmentalism is viewed as an important element of good citizenship, that general criticism seems valid here.[41]

Despite an admitted degree of uncertainty, however, the Gallup data can be defended for two reasons. First, results indicate only what individuals and their family members have done "in the past year," not what activities they regularly participate in. In this sense, high frequencies that accumulate behavior over an extended period of time are not unexpected. Moreover, even if participation is overreported by eager respondents, the relative rank of activities (similar to that of environmental concern and issue salience) remains less affected. And it is, in fact, that rank that leads to the most compelling and transparent conclusion of all. Quite simply, Americans are more like to act "green" when there

Table 2.4

Self-Reported Participation in Thirteen Environmentally-Responsible Activities

Which of the following, if any, have you, yourself, done in the past year?

Question	Yes, have done
Voluntarily recycled newspapers, glass, aluminum, motor oil, or other items	90%
Avoided using certain products that harm the environment	83
Tried to use less water in your household	83
Reduced your household's use of energy	83
Bought some product specifically because you thought it was better for the environment than competing products	73
Contributed money to an environmental, conservation, or wildlife preservation group	40
Signed a petition supporting an environmental group or some environmental-protection effort	31
Voted for or worked for candidates because of their position on environmental issues	28
Attended a meeting concerning the environment	20
Contacted a public official about an environmental issue	18
Been active in a group or organization that works to protect the environment	15
Contacted a business to complain about its products or policies because they harm the environment	13
Bought or sold stocks based on the environment record of the companies	9

Source: Gallup Organization (April 3–9, 2000).
Note: $n = 1,004$.

is a tangible incentive to do so and when the need for personal sacrifice is slight.[42]

For example, the most frequent task reported to Gallup involved recycling "newspapers, glass, aluminum, motor oil, or other items." Nearly all of those interviewed (a staggering 90 percent) reported that they had recycled those products within the past few years, but it is important to keep in mind that recycling is now mandatory in many cities and municipalities, often with curbside pick-up—a basic fact that stands at odds with Gallup's use of the term "voluntarily."[43] As some scholars point out, policies such as this are a form of forced behavior change.[44] In other states, such as Maine and Vermont, consumers are reimbursed a small deposit fee after bottles are emptied and returned to community redemption centers. Some families, of course, recycle household waste out of a genuine concern for the environment or an intrinsic sense of personal satisfaction,[45] but others undoubtedly do so because of a financial incentive or a need to comply with local ordinances.[46] Both are clearly factors that suggest that recycling cannot be seen as a full reflection of the public's commitment to environmental protection.

Notice, too, that the Gallup survey asked respondents about whether they had "tried to use less water" or whether they had reduced their household's use of energy. The motivation of respondents in performing these tasks is equally unclear. As efforts that conserve natural resources, both could be considered environmentally friendly, and yet the financial benefits that respondents receive by reducing the costs of household utilities are not trivial. In the end, it seems fair to say that environmentalism in practice is frequently a by-product of activities that are justified by other rewards. Faced with rising expenses, Americans have learned that it often pays to be green. But if such is the case, environmentally responsible behavior may depend on the presence (or absence) of those cues, signaling little about attitude strength and conviction.

Understanding Ambivalence

Given the challenges of survey research, the argument presented in this chapter so far has been necessarily circuitous, and yet data from a wide variety of sources using different formats and techniques all seem to point to the same basic conclusion. High expressions of support notwithstanding, actual public commitment to the environment is limited, especially when personal sacrifice and competing priorities are brought to mind. In this respect, of course, the environment is truly no different than any other political issue. Public support is never unconditional, nor are Americans ever anxious to assume the financial burden of their demands, even on issues they care about.[47] Yet in commenting on recent

trends in environmental conditions, Ed Ayres of the Worldwatch Institute writes, when "plotted on graphs, they look like heart attacks," and
so, he wonders, "why are we *not* astonished?"[48] Why do Americans not
express environmental concern with greater depth and resolve, especially
given their belief that these problems are both real and dangerous?

First, perhaps average citizens fail to take a more intense personal
interest in the environment because they trust that those problems are
being satisfactorily addressed by others, including policymakers and
business leaders. Mitchell finds this explanation entirely "plausible"
given increased government funding for environmental regulation since
1970.[49] In general, however, while levels of satisfaction with government
involvement on the environment have risen somewhat over the past de-
cade, many Americans remain cautious and pessimistic.[50] For example,
in the April 2000 Gallup poll cited extensively in this chapter, 64 percent
of those polled believed that "only some progress" had been made in
dealing with environmental problems, while 60 percent had "only some
optimism" that those issues would be "well under control" within the
next twenty years.[51] In this context, the seeming intractability of environmental problems may be to blame. As Mitchell argues, people often
feel that there is "nothing they can do about the problem despite the
strength of their feelings,"[52] a factor that can lead to resignation rather
than forward momentum.[53]

Second, perhaps environmental issues are too far removed from personal experience to motivate greater depth of concern.[54] As Philip Shabecoff reasons, "The environment is not usually an issue of high political
salience, but when the quality of their air and water and the health of
their children are threatened, Americans can be roused to anger."[55] Data
available in the 1995 pilot of the National Election Study (NES), however, find surprisingly little support for that logic. Concern for local air
quality may indeed motivate Americans to support government efforts to
combat air pollution, but respondents overall were far more likely to
complain about air and water quality nationwide than they were about
their own communities, as table 2.5 shows.[56] That result finds additional
support in Robert Rohrschneider's work cross-nationally, which shows
that self-interest plays a diminished role in understanding public attitudes on the environment.[56]

Table 2.5
National vs. Local Differences in Environmental Concern

Overall, how would you rate the air quality in ...

Response	Our nation	Your local community
Very good	5.9%	29.1%
Fairly good	59.7	48.6
Fairly bad	26.9	15.1
Very bad	7.6	7.2

Source: National Election Pilot Study (1995).

Overall, how would you rate the safety of drinking water in ...

Response	Our nation	Your local community
Very good	11.2%	34.8%
Fairly good	56.2	44.3
Fairly bad	26.0	14.5
Very bad	6.6	6.4

Source: National Election Pilot Study (1995).

Finally, given that Americans tend to place priority on immediate concerns over long-term uncertainties, an added cause of weak attitude strength may be the distant time horizon perceived on many environmental threats, such as global warming and ozone depletion. This interesting hypothesis receives some support in table 2.6. For instance, as noted earlier, just 2 percent of those responding in the Gallup study cited the environment or related pollution concerns as one of the nation's "most important problems," and only 1 percent did the same in reference to their own community. When attention is directed twenty-five years into the future, however, 14 percent of those polled volunteered environmental concerns, pushing the environment to the top of the list. In other words, as David Helvarg notes, "environmentalism is, if not moving to the front burner, at least heating up the back of the stove."[57]

Conclusions

To say that environmental issues are embraced by a majority of Americans is hardly a controversial statement. Scholars have long noted

Table 2.6
Ranking Important Problems: Locally, Nationally and in the Future

What do you think is the most important problem facing ...

Issue	Our nation today	Your community today	Our nation 25 years from now
Environment/pollution	2%	1%	14%
Crime/violence/guns	17	19	7
Decline in morals/ethics	13	3	8
Education/schools	11	17	6
Dissatisfaction with government	11	2	2
Economic concerns	10	12	9
Health care	6	3	3
Poverty/homelessness	6	1	3
Drugs/alcohol abuse	5	10	3

Source: Gallup Organization (May 23–24, 2000) ($n = 1,032$); and April 3–9, 2000 ($n = 1,004$).
Note: All responses are open-ended. Question wording for community concerns included the phrase "worst problem" rather than "most important problem." Items in list do not add to 100 because some issues with a small number of responses were excluded.

that the environment has become a consensual issue, perhaps to the point of defining a new cultural norm.[58] There are few obvious anti-environmentalists today, and no major bloc of voters opposed to environmental protection.[59] Yet where we might presume black and white there are, in reality, only shades of gray. The real question is no longer whether Americans side with environmentalism but rather what kind of commitment they bring to the table.

Unfortunately, as the introductory chapters caution, the power of overwhelming numbers is not always what it seems. Diffused by survey research that asks the right questions, public resolve on the environment seems weakened by ambivalence, conflict, and contradiction, especially when the inherent social desirability of those issues is countered by other equally important goals in ways that imitate political reality. This conclusion is consistent with the way in which a majority of Americans describe themselves in polls—as "sympathetic, but not active" within the environmental movement.[60]

In the end, that fact should not be ignored, either by survey researchers or by political activists who seek to use polling results to pressure government and business leaders into forming protective environmental policies. "In the United States," writes Mark Dowie, "ecology is a household word and almost everyone is an environmentalist," but a consensus based on words alone may prove to be shallow and transient in the long run, endangering the legitimacy and political base of environmentalists who place too much faith in public mandates and grassroots support.[61]

3
Stability: Have Environmental Attitudes Changed over Time?

In a well-known article titled "Up and Down with Ecology," published in 1972, just two years after the first observance of Earth Day, Anthony Downs compared public opinion on the environment to a broader concept he labeled the issue-attention cycle. For Downs, it was a pattern defined by the mercurial tendency of Americans to shift notice from one problem to the next. "Public attention rarely remains sharply focused upon any one domestic issue for very long," he said, "even if it involves a continuing problem of crucial importance to society."[1] Instead, Downs cautioned that most issues populating the public agenda follow a predictable and nearly inevitable orbit that is marked initially by "alarmed discovery" and "euphoric enthusiasm" but later by the reality of rising policy costs, which brings with it discouragement and declining support.[2] Declaring that public interest in the environment was already midway through that cycle, his message to environmentalists was clear. It might be "possible to accomplish some significant improvements in environmental quality," but only if "those seeking them work fast."[3]

Given the wisdom of hindsight, of course, Downs's forecast seems decidedly mistaken or at the very least premature. Interest in environmental issues has not fallen off the political map but in truth it has not remained constant and unwavering either, at times shifting around definitive peaks and troughs. For example, in the early 1970s Hazel Erskine acclaimed "a miracle of public opinion," writing that "ecological issues have burst into American consciousness" with "unprecedented speed and urgency."[4] Within a few short years, however, most scholars confirmed a rapid retreat.[5] A renewal of environmental concern in the late-1980s followed (one that to some constituted a "second miracle"),[6] but just a few years later poll watchers pointed to evidence of decline, noting that "acute concern" had eroded once again and that environmental issues had progressed to the midpoint of yet another life cycle.[7]

Within this fluctuation there is, perhaps, an inference to be drawn. If trends in support for environmental protection periodically decline for extended lengths of time (for any number of imaginable reasons), environmentalism may be little more than a political fad or fashion subject to the nature-of-the-times. We might suspect that near consensus on environmental issues is due, possibly, to the fiscal generosity that accompanies economic prosperity or to the subliminal message of media agenda setting rather than to honest public concern.[8]

In part, those doubts seem reasonable in light of scholars' long-standing tendency to assess the quality of American public opinion by wider reference to the characteristics described by Downs—to attributes that, in V. O. Key's words, "speak of the volatility of public opinion, its whimsicality, its sluggishness, its stability, its erraticism, and its unpredictability."[9] Indeed, the very definition of *attitude* found in most textbooks requires that citizens develop a "learned predisposition to respond" that is largely consistent and stable when subject to scientific measurement.[10] Whether a majority of Americans meet that threshold on issues like the environment is frequently challenged. While some complain that opinion change over time consists of little more than white noise,[11] others counter that policy preferences are remarkably stable, characterized by change that is slow, steady, and, most important of all, explicable.[12]

Addressing the accuracy of these competing claims forms the basis of this chapter. In concentrating on the stability of opinion, it is a plan that moves focus to a broad landscape, one described not simply by cross-sectional slices of public opinion at static moments in time but rather by dynamic shifts within an aggregated public *across* time. As a result, an appropriate goal is twofold—to categorize the nature of change in environmental attitudes and then to explain deviations from that course in a powerful but ultimately parsimonious model that uses four sets of variables—cohort replacement, economic conditions, federal spending, and media attention.

Tracking Trends in Environmental Attitudes

To study the aggregate stability of public opinion requires that at least two elements be in place.[13] First, trends in identically worded questions asked by major polling organizations must be identified and compiled (no small task in a profession that at times prizes responsiveness to new events over long-term continuity). This criterion is necessary to ensure that change in attitude alone is measured and not some lesser variation in question format, wording, or design.[14]

Second, an accompanying theory of change must be defined and tested against the evidence. As D. Garth Taylor writes, on that count there seem to be several possible models.[15] For example, under some conditions public opinion might be *constant*. That is, it might exhibit little real change, where differences from one datum to the next are not statistically different from zero. In analyzing 137 attitudinal trends found in the NORC General Social Survey (GSS), Tom Smith finds that 34 percent of issues qualify under those conditions.[16]

A second model identifies a pattern of *linear* change that either increases or decreases over time at a slow, steady, and reliable rate, often due to cohort replacement or to broader structural changes such as educational attainment or labor-force participation. According to Smith, 41 percent of GSS trends demonstrate either clear linearity or a significant linear component (one where unexplained variance is still present). Two final theories recognize the potential of short-term, nonlinear change. In some cases, shifts in opinion might be *cyclical*, paired with repetitive market trends or election schedules.[17] In others, change might be *episodic*, charting sudden and erratic shifts based on events of unique public significance, such as war, assassination, or depression.[18]

Each of these models seems intuitively plausible when applied to the environment. The natural replacement of one birth cohort with another might certainly create linear growth in environmental concern, particularly among generations first socialized to the dangers of environmental degradation in the years since 1970.[19] Cyclical change might likewise be produced by any number of factors, including economic conditions like unemployment or inflation or even the scientific discovery of new environmental problems, where progress in some areas is quickly offset by the emergence of other pressing concerns, allowing Downs's issue-attention cycle to repeat itself time and again.[20] Finally, event-driven models might produce opinion *bounce* due to environmental disasters that are well publicized by the press, such as those that occurred at Love Canal, Three Mile Island, Bhopal, Chernobyl, or Prince Edward Sound in 1989, where massive quantities of oil were spilled by the *Exxon Valdez*.

To identify which model (or combination of models) is best suited to environmental attitudes, long-term trends in five survey questions—all composed of at least ten observations each between 1973 and 1999—are graphed in figure 3.1.[21] Chief among them is a popular measure of support for spending on "improving and protecting the environment" asked annually, with few exceptions, in the NORC General Social Survey (GSS). Despite differences in question wording and intent, those five trends seem to confirm a common result. Interest in environmental protection declined or remained stagnant in the 1970s, rose to record levels between 1987 and 1991, and then descended again in the years that followed, corroborating the work of many other observers in the field.[22]

But how might that visual conclusion be formalized by way of the theoretical models outlined earlier? That question is easier to answer when focusing on one trend at a time, most appropriately the longest trend available, which is the GSS spending measure. A close look at that data shows that the magnitude of change in the percentage of respondents who believed that "too little" was being spent on the environment in each year in which the question was asked is small, even incremental. Movement on a wider scale is evident, however, when successive observations of that variable are regressed against time, as shown in figure 3.2. The graphical display there suggests that an underlying linear trend drives public attitudes on the environment, with a rate of change of nearly four percentage points per annum.[23] Yet even more apparent are cyclical deviations from that trend that remain unexplained in estimating the equation.[24] Indeed, this cycle moves with remarkable regularity, deviated by no obvious outliers that might point instead to forces that

Percent responding

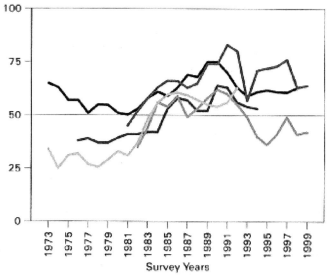

Survey Years

▬ Government spends "too little" on the environment.
▬ We must sacrifice economic growth for environmental protection.
▬ We must make environmental improvements "regardless of cost."
▬ Government regulation on the environment is "too little."
▬ Environmental laws and regulations have not gone far enough.

Figure 3.1
Trends in Support for Environmental Protection, 1973 to 1999
Note: All missing values have been interpolated in order to provide an uninterrupted time series. See appendix for question wording.

are random and erratic. Studying that duel pattern further requires a more sophisticated model, as well as a clearer sense of causality.

Explaining Opinion Change

V. O. Key writes that "stability of opinion both in idea and in fact can be understood only in relation to the stimuli" that affect it.[25] That conclusion, for him, drew natural attention to a series of issues, problems, and cues, the most promising of which for present purposes are cohort replacement, economic conditions, federal spending, and media attention.

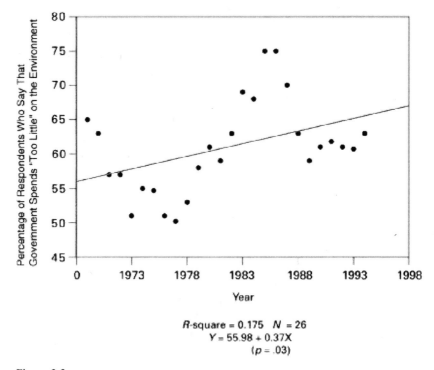

$$R\text{-square} = 0.175 \quad N = 26$$
$$Y = 55.98 + 0.37X$$
$$(p = .03)$$

Figure 3.2
Deviations from a Linear Trend in Support of Increased Environmental Spending, 1973 to 1998
Source: NORC General Social Survey (various years).
Note: Missing values for the years 1979, 1981, 1992, 1995, and 1997 have been interpolated to create an uninterrupted time series.

Cohort Replacement

First and foremost it is important to recognize that attitude change is likely to come from two sources. In the aggregate it can be bought on by the shifting preferences of individuals (whose motivations are explored below), or it can result from gradual cohort replacement. Specifically, the latter refers to that part of a trend that is due to movement along a cohort's birth-year axis.[26] In this sense, we might predict linear growth in environmental concern over time—consistent with figure 3.2—as successive generations of older, less environmentally aware citizens are replaced by those who are younger and more conscious of environmental degradation, especially in the years following the creation of a modern environmental movement.[27]

Economic Conditions

Scholars have long argued, too, that policy preferences are shaped over time by economic conditions and expectations.[28] In fact, on the environment there is evidence that public concern is conditioned by the business cycle, rising in social priority during periods of prosperity and falling when times turn hard.[29] As one journalist explains from the vantage point of early 1992,

When times get tough, the questions facing environmentalists get even tougher. And these days, economic anxieties and shifting political winds are threatening to produce a green-out effect that could make tree huggers feel as endangered as the California condor. Epochal events such as the Gulf War and the collapse of the Soviet Union have pushed most domestic ecological concerns off the front pages. The recession has prompted many people to question the costs of environmentalism and made it harder for preservation groups to raise money and boost membership. In the presidential campaign, saving the planet has become an orphaned issue. No savvy candidate would dwell on ozone depletion and the need for biodiversity when voters are worrying about whether they'll have a job next year or be able to pay their medical bills.[30]

Based on that advice, it is not surprising that then-incumbent George H. W. Bush—who just four years earlier in better times had pledged to be the "environmental president"—decided that an election year turnabout was needed, telling voters on the campaign trail that year that it was now time to "put people ahead of owls."[31]

According to Euel Elliott, James Regens, and Barry Seldon, downturns in the economy are potentially devastating for environmentalists in pursuit of political goals, even more so given that perceptions of economic conditions are often as important as the statistics themselves.[32] They argue that "to the degree economic growth is modest, hard-won gains obtained by the environmental movement, as well as advances in environmental protection, may be in serious jeopardy if implicit trade-offs between economic performance and environmental quality dominate the policy agenda."[33]

Federal Spending

A third possible factor explaining temporal shifts in environmental opinion associates those trends with satisfaction with current efforts and expense. As chapter 2 notes, once public outrage succeeds in inducing government to assume responsibility for improving environmental quality, citizens may be likely to feel a sense of satisfaction in the belief that the problem "is being taken care of."[34] In response to actual government spending, therefore, Christopher Wlezien believes that the public behaves much like a thermostat, adjusting "its preferences for 'more' or 'less' policy in response to policy outputs themselves."[35]

Media Attention

Finally, public responsiveness to external factors like economic conditions and federal spending is communicated (and to some extent mediated for) the public by the news media.[36] As Benjamin Page and his colleagues write, when it comes to acquiring political information "most people rely heavily upon the cheapest and most accessible source" available to them, including television, radio, newspapers, and magazines.[37] That dependence is significant here because studies show that media coverage on the environment tends to be event oriented, rising and declining in a pattern that closely matches the public mood, a fact that has encouraged some scholars to go still further in attributing cause. For example, in summarizing published scholarship in the field, Stuart Allan, Barbara Adam, and Cynthia Carter note that concern for unemployment and inflation displaced environmental issues from the news agenda in the 1970s, helping to precipitate a rapid decline in policy support among average Americans "once the reinforcing and sustaining influence of the mass media disappeared."[38]

In Search of a Model

Our primary interest here is to understand why environmental attitudes change over time—specifically, why aggregate changes occur in the percentage of those who believe that "too little" is being spent on "improving and protecting the environment," as measured annually by the NORC General Social Survey (GSS) between 1973 and 1998.[39] In the language of basic statistics (as in algebra), that pooled set of data is considered a dependent variable—that is, a variable whose value is determined by the presence or degree of one or more independent variables, which in this case include those described above (cohort replacement, economic conditions, federal spending, and media attention).

Incorporating four highly reasonable expectations of opinion change into a formal model subject to empirical testing, however, is not as easy as it might appear, but it is necessary if the relative contributions of each are to be isolated and disentangled. The approach used to accomplish that goal, called *multivariate regression analysis*, is popular in many disciplines, and is accessible to most readers with at least some statistical training.[40] An added wrinkle must be dealt with here, however. Since the pooled GSS data construct a time-series, the problem of *autocorrelation* must be confronted before estimating any model. This term is used by scholars to refer to a condition in which the error terms corresponding to different points in time are correlated with one another. As basic econometrics textbooks are quick to point out, failing to acknowledge a lack of independence can lead to an overestimation of regression coefficients and of a model's explanatory strength.[41]

One possible corrective for autocorrelation is to take first differences of variables on both sides of the equation. By subtracting each datum in the series from its predecessor, the series is detrended and made stationary. According to Gary King, however, that approach can "cancel out" important systematic components in the data, so much so that "models based on differenced series tend to fit less and have higher standard errors and less stable coefficient values."[42] Given those concerns, Robert Durr argues that "prewhitening" techniques such as these can be akin to "throwing out the baby with the bathwater," especially since alternatives are widely available.[44] As he recommends, empirical results are generated instead using a first-order autoregressive error model found in the statistical program SAS, the results of which produce a fully acceptable Durbin-Watson statistic.[45]

One last set of difficult decisions involves the operationalization and measurement of the independent variables used in the equation. Cohort replacement, for instance, is a complex process not easily reflected in an aggregate time-series.[45] Still, the possibility is explored here by including the percentage of GSS respondents born after 1945 as an independent (or explanatory) variable in a multivariate regression—a technique that is explained in greater detail below. As might be expected, that proportion increases over time in a linear fashion as younger birth cohorts, socialized to the significance of the environmental movement, are gradually added to the survey population.

The remaining variables in the equation all attempt to capture cyclical or episodic deviations from that underlying trend, including a dummy variable that offers statistical control for the presence of Democratic or Republican presidential administrations. Two measures representing objective economic conditions widely reported to the public (annual measures of unemployment and inflation) are also included, as is one subjective factor, the Conference Board's Index of Consumer Confidence. Constructed from a survey of 5,000 households, the latter is an important measure of the public's mood on business conditions and the labor market.

Finding an appropriate indicator of government spending on the environment presented an especially difficult challenge.[46] Two possible statistical series were rejected for use here. One, preferred by Euel Elliott and his colleagues, is the Pollution Abatement Costs and Expenditures survey (PACE), but its selective focus suggests that it is too narrow for our present purposes.[47] The PACE index is casually flawed for one additional reason as well. It primarily tracks capital and operating costs incurred by manufacturers in the private sector, while the question asked by the General Social Survey implies a political and social choice better represented by government spending. A second plausible measure of spending on the environment is found in the U.S. Environmental Protection Agency's (EPA) extensive report, *Environmental Investments: The Cost of a Clean Environment*, published in 1990.[48] Like the PACE survey, however, measures developed in that report are limited to pollution-control costs. Moreover, given its date of publication, estimates for the years 1990 through 2000 are not firm but rather are linearly extrapolated based on shifting assumptions about a variety of implementation scenarios. To avoid these difficulties, federal budget expenditures itemized under the broad category natural resources and the environment are used here instead. Those figures are adjusted for inflation and lagged one year on the understanding that some length of time is necessary for alterations in spending priorities to be noted by a lay public.

Finally, since no content analysis of media attention to environmental issues could be found extending backward in time nearly thirty continuous years, one was created for this purpose by counting the number of news stories indexed each year under the headings Environmental Policy and Environmental Movement by the *Readers' Guide to Periodical Literature*. That number ranges in value from a low of forty-eight articles in 1979 to a high of 240 in 1992.

Model Estimation

With several concerns now resolved, final data results are reported in table 3.1. As expected, the gradual shift of birth cohorts has a positive, linear impact on support for environmental spending over time, confirming the intuition of Samuel Hays and others. Yet it is important to note that generational change remains a force challenged and at times offset by movement in other arenas. For example, it is clear that economic conditions dampen public enthusiasm for policy efforts.[49] Holding all else constant, a one percentage point increase in the annual rate of unemployment depresses public support by more than four percentage points. Put another way, the isolated difference in public opinion between a period of high unemployment (1982) and one of low (1998) could be twenty percentage points or more—certainly a large enough shift to exert pressure on the spending priorities of Washington lawmakers.[50]

Second, while federal budget expenditures are a statistically significant predictor of public attitudes toward spending on the environment, its impact is surprisingly small, even with hypothetically large increases in funding. The failure of that "thermostat" to register more clearly may well be a consequence of imperfect information. "After all," as Wlezien notes, "there is reason to think that specific information about appropriations for these programs is not regularly and widely available to the public."[51] More powerful, it seems, is the partisan identity of presiden-

Table 3.1
Autoregressive Model Explaining Trends in Support of Increased Environmental
Spending, 1973 to 1998

[Dependent Variable:] *We are faced with many problems in this country, none of which can be solved easily or inexpensively. I'm going to name some of these problems, and for each one I'd like you to tell me whether you think we're spending too much money on it, too little money, or about the right amount on ... improving and protecting the environment.*

Independent variables	Yule-Walker estimate	Standard error
Percentage of GSS respondents born after 1945	0.744**	0.222
Unemployment rate, seasonally adjusted, expressed as a percentage of the total labor force	−4.485***	0.952
Rate of inflation, as measured by percentage change in the Consumer Price Index	−0.889*	0.378
Index of Consumer Confidence (1985 = 100)	−0.077	0.058
Federal budget outlays for natural resources and the environment, lagged one year (in millions of 1996 dollars)	−0.002**	0.001
Dummy variable for presidential administration (0 = Democrat, 1 = Republican)	6.291**	1.885
Number of news media articles on environmental policy and the environmental movement indexed by the *Readers' Guide to Periodical Literature*	0.040*	0.018
Intercept	78.642	13.560

R-square = 0.88
Durbin-Watson = 2.06 (probability = 0.13)
Number of cases = 26

Sources: NORC General Social Survey (various years).
Notes: Missing values in the dependent variable for the years 1979, 1981, 1992, 1995, and 1997 were interpolated to provide an uninterrupted time series.
$*p < .05.$
$**p < .01.$
$***p < .001.$

tial administrations, which may serve a comparable function using far simpler means by providing the public with a shorthand cue to the incumbent's ideological commitment to environmental protection. Net of other effects, results there suggest that respondents are far more likely to believe that "too little" is being spent on the environment when Republican presidents are in charge—a perception (interestingly enough) that is not always borne out in fact.[52]

Third, data results indicate that media attention to environmental policy and the environmental movement itself can produce small-scale changes in public attitudes, most likely through the power and influence of agenda setting. Given the tendency of news coverage to follow the ebb and flow of major environmental disasters like Chernobyl and the *Exxon Valdez*, it is not difficult to imagine a scenario in which an additional one hundred articles published in major news magazines in a given year increase public support, holding all else constant, by four percentage points.[53]

One final conclusion on a broader scale also seems warranted. Given the overall fit of the model and its ability to explain an overwhelming proportion of the variance observed using relatively few regressors, the data speak to lingering controversy over the stability of mass attitudes. While some scholars continue to view opinion as "unreliable chaff" with little "substantial grain to measure," results here confirm instead the sentiment of Tom Smith, who argues that most change is intelligible over time, driven by events of social, political, and economic importance. Indeed, as he writes, there seems to be "neither chaos nor a chimera, but rather order and a map of reality."[54]

Conclusions

Most early studies of environmental attitudes found the "social mood of consensus" following the first Earth Day in 1970 to be "substantially superficial," leading scholars to predict "a rapid, general retreat from hard-nosed environmental reform" as soon as the costs of conservation and regulation became apparent. Others, too, argued that under the pressure of a struggling economy and a serious energy crisis, concern for environmental issues would inevitably (and understandably) weaken. The basic facts that underscore each of those arguments receive solid support in this chapter.

Just as important, however, is the often overlooked ability of public opinion to find rejuvenation. Unexpected environmental events and disasters reported by the news media, as well as the continued discovery of new sources of environmental degradation by scientists, likely has helped to sustain long-term interest, allowing the environment to avoid the policy fate of Downs's issue-attention cycle by recharging public attention and enthusiasm time and again.[55]

Still, a pattern of attitudinal peaks and troughs over the past thirty years suggests that environmental concern remains conditional, rising and falling in response to economic conditions, policy costs, media attention, or even outright public boredom. Within that cycle an unspoken assumption also exists—namely, that public support for the environment is more a political fad or fashion subject to the nature-of-the-times, than it is a valued partner in the environmental movement. Yet to reach that far in a normative direction based on evidence of opinion change alone seems premature for several reasons.

First, as James Stimson argues, shifting policy preferences on a wide variety of issues tend to move together over time, closely following the undulations in environmental attitudes observed here.[56] In other words, the factors that influence public opinion on the environment do not exist in a political or social vacuum. Support for new policy initiatives may weaken during economic downturns, but it appears to do so across the board. As such, temporal shifts need not indicate a change in the *relative* priority placed by the public on environmental protection.[57]

Finally, the movement of environmental attitudes in response to exogenous forces may worry environmentalists intent on preserving popular support, but the order and predictability of those changes compensate, in part, by highlighting well-defined "policy windows" or moments that are more (or less) ripe for opportunity.[58] As Robert Durr notes, that knowledge "offers considerable strategic guidance" to lobbyists and lawmakers alike,[59] allowing savvy environmentalists to maximize political gains during periods of high, salient support and to brace areas of vulnerability during less advantageous times.

4
Distribution: Is Environmentalism Elitist?

Fluctuating trends in environmental concern over the past thirty years have led to a growing recognition that environmental issues and economic interests remain intertwined, increasingly (and uncomfortably) so on matters of social justice.[1] For example, some argue that communities of color are disproportionately burdened by environmental risk in the form of landfills and incinerators.[2] Others say that a concern with urban sprawl and open space reflects the "selfish downside of democracy," where decisions are made by a middle class intent on preserving its own quality of life at the expense of others through the use of exclusionary zoning.[3]

The potential for conflicts such as these to unleash "thorny issues" of elitism seems very real, especially if average Americans view environmental issues differently based on their social, political, and economic standing.[4] As Philip Shabecoff warns in *Earth Rising* (2000), a weight of tradition within the environmental movement has made it reluctant "to embed itself in the workday human community—where people live, worry about their jobs, send their children to school, go to church and synagogue and mosque, and are exposed to myriad social as well as physical insults in their environment." As a result, the popular image of an environmentalist remains that of a "backpacker and a tree-hugger," someone concerned mainly with wildlife protection and wilderness preservation.[5] It is a dogmatic perception that seems to invite little common ground.

In contrast, however, some argue that in recent years pressure exerted on the movement from the bottom up has succeeded in bringing "Earth

Day back down to earth."[6] As Mark Sagoff maintains, today's environmentalism "serves as a common rallying ground for groups usually thought to be at odds with one another: educated professionals and the lower middle class; affluent suburbanites and inhabitants of small towns in the American heartland."[7] For these reasons, he believes that a more inclusive environmental agenda has spanned the chasm between a society of "haves" and "have nots," creating an entirely new breed of populism in the process.

Deciding which of those two competing views is more accurate in the end remains open to debate but ironically enough, not for lack of effort. Understanding the dynamics of environmental concern by reference to its political and demographic base has long been a challenge for scholars. In fact, it would not be unfair to suppose that in the field of public opinion on the environment, no single issue has generated as much energy or attention. Yet in spite of that research, empirical results have been surprisingly mixed. While some studies confirm that environmental values and beliefs lead to familiar cleavages based on age, education, income, gender, and partisanship,[8] others argue that polarization has been greatly exaggerated, pointing out that those factors (in sum) rarely account for more than 10 to 15 percent of the variance observed.[9]

Ironically, a multitude of contradictions and discrepancies in the field seem best explained by reference to a now familiar problem. Careful comparisons between results and also methodologies appear to indicate that the importance of demographic predictors varies, depending heavily on the way in which environmental issues are framed in questionnaires.[10] Without keen attention, then, to issues of measurement and to theories about underlying causes, it should come as no surprise that numbers alone "do little to clarify the picture" and in some respects serve only to "muddy it" further.[14]

This chapter intends to revisit long-standing charges of environmental elitism through the corrective lens of those two issues.[12] The first step involves clarifying theoretical expectations regarding various social, political, and economic groups toward the environment. The second step involves formalizing those relationships in a series of multivariate regression models that allow the influence of each group trait to be isolated from those that remain.[13]

Demographic Expectations

If it is reasonable to suppose that environmental issues are interpreted and prioritized in different ways by different people, what cleavages predominate? Six political and demographic predictors, in particular, have received extensive attention in published scholarship on environmental attitudes—age, education, income, race, gender, and partisanship.

Age

A respondent's age has been called the "strongest and most consistent predictor of environmental concern" for reasons solidly grounded in theories of opinion formation and value change.[14] As Samuel Hays explains in *Beauty, Health, and Permanence* (1987), the age variable matters because it relates to differences in the shared life experiences of birth cohorts.[15] Like Hays, studies here often emphasize the ripe development of younger ecology-minded generations since 1970—those socialized into a world newly shaped by a recognition of environmental degradation and "limits to growth." In this sense, the rise of environmentalism joins, in Ronald Inglehart's words, a "post-materialist" agenda focused on quality-of-life concerns.[19] Spurred by satisfaction with economic well-being, this trend is manifested slowly in polls by cohort replacement as one generation succeeds another in adult society.[17]

Education

Expectations regarding educational differences in environmental attitudes are also clear in the literature. According to many scholars, higher levels of education increase cognitive skills and support an awareness of public affairs, both of which give individuals a greater interest in and ability to comprehend complex environmental problems.[18] Respondents with higher levels of education also seem better able to translate concern into effective action because of an availability of resources, the likelihood of organizational affiliations, or a heightened sense of personal and political efficacy.[19]

Income

Measures of income and wealth constitute a third possible determinant of environmental concern. Charges of elitism, after all, have remained a constant theme among critics of the U.S. environmental movement over the past thirty years.[20] With its strong attack on unbridled economic growth and the exploitation of natural resources beyond the limits of sustainability, environmentalism has led some to argue that the movement harbors an inherent asceticism that is unresponsive to the economic needs of those less privileged.[21] This belief finds justification in Inglehart's theory of postmaterialism and in A. H. Maslow's hierarchy of needs, both of which suppose that concern for higher-order issues like the environment develops only after basic economic needs are satisfied.[22] To put it bluntly: "Struggling inner-city communities rarely have had the time, or the inclination, to save the whales."[23]

Race

Fourth, low participation of minorities in the environmental movement and a suspected "concern gap" between black and white respondents have led some researchers to label environmentalism a "secular religion of the white middle class."[24] As Robert Bullard contends in *Dumping in Dixie* (1990), mainstream environmental groups have tended to stress nonhuman issues such as wilderness and wildlife preservation and resource conservation, which are not issues of high priority to communities of color, particularly those inhabited by low-income urban residents with little leisure time and inadequate transportation to recreational areas.[25] Although environmental health threats would seem to affect all urban poor (black and white alike), some argue that minority groups suffer from institutional barriers that block residential mobility, including discrimination in local housing markets.[26]

Gender

Research on gender as a predictor of environmental concern represents another prodigious area of research.[27] Many scholars suggest that women are more environmentally concerned than men based on their maternal socialization as family nurturers and care givers.[28] Others, too, argue that women tend to feel a heightened vulnerability to environmental risk,[29] especially when attention is focused on new technologies,[30] or on local hazards that impact public health and safety, such as toxic waste disposal and nuclear power.[31] Moreover, given that many women retain a traditional role in assuming household duties and making consumptive decisions for their families, gender differences also seem apparent when examining willingness to purchase environmentally responsible products in the marketplace.

Partisanship

Finally, critics have long argued that environmental priorities remain firmly entrenched to the left of a "deep ideological divide" that separates environmentalism from a commitment to personal liberty, property rights, and technological optimism.[32] It is not surprising, then, to find that political ideology and partisan identification are each a "highly consistent predictor" of environmental concern.[33] After all, environmental regulation is commonly opposed by political conservatives who object to its financial cost and to an expansion of government interference in the market economy. Many, too, remain cautious about expansions of federal and state bureaucracy that oversee and complicate private land-use decisions or create additional layers of "red tape." For all of these reasons, scholars expect environmental attitudes to be closely associated with liberal Democratic views.

The Measurement Quandary

Conventional wisdom and theoretical expectations notwithstanding, the empirical record on political and demographic predictors of environmental concern is mixed. Kent Van Liere and Riley Dunlap have "confidence in concluding that younger, well-educated, and politically liberal persons tend to be more concerned about environmental quality than their older, less educated, and politically conservative counterparts."[34] Yet given an accumulation of weak results, that degree of assurance is rarely shared by others, one of whom notes that the field remains an "unsettled research area."[35]

Why should the correlates of environmental concern appear inconsistent across studies? The primary reason is methodological, according to Stephen Klineberg, Matthew McKeever, and Bert Rothenbach, who argue that "the determinants of environmental concern vary greatly depending on the wording and framing of the questionnaire itself."[36]

Theirs is a reasonable explanation that appears even more likely when a full menu of environmental measures used in surveys is considered. For example, while some scholars look to public perceptions of the seriousness of environmental problems,[37] others prefer support for government spending,[38] knowledge about environmental issues,[39] environmental policy preferences,[40] or self-reported involvement in environmentally responsible activities, such as recycling or energy conservation.[41] They argue that each of these measures introduces different costs and trade-offs into the minds of survey participants, stimulating inconsistent responses. Questions focused on government policy, including issues like taxes and spending, are likely to confound environmental concern with political ideology. An emphasis on voting and volunteer work might confuse environmental support with a more general and obvious relationship between socioeconomic status and political participation.[42] A reliance on marketplace activities could privilege income variables, confusing willingness-to-pay with the ability to do so. For all of these reasons, it should come as no surprise that it may matter greatly "how environmental concern is measured."[43]

A "New Class" or a New Consensus?

When concerns about the environment are being surveyed, careful attention must be paid both to the theories that underscore the utility of certain independent (or explanatory) variables like income, education, and partisanship and also (as Klineberg and his colleagues suggest) to the way in which dependent variables are framed. With those joint issues in mind, the analysis that follows revisits the long-standing debate over the political and demographic bases of environmental concern by comparative reference to four different factors—perceptions of environmental danger, environmental policy preferences, strength of feeling toward environmentalists, and self-reported participation in environmentally responsible activities.

Perceptions of Environmental Danger

As noted previously in chapter 2, the most elementary way of measuring public attitudes toward environmental protection is to ask citizens to evaluate the depth of their feelings directly, often using a scaled response format. This approach was used in the 1994 NORC General Social Survey (GSS), where respondents were asked how dangerous they thought a number of environmental problems were "in general" and for "you and your family," including air and water pollution, nuclear power, pesticide use, and the greenhouse effect.[44] For present purposes, responses to the first set of questions were summed into an additive scale of increasing environmental concern ranging in value from six to thirty points. Responses to the second set of questions were also summed into an additive scale of equal length, focusing instead on personal perceptions of environmental danger. As in chapter 3, the effects of a series of demographic and political variables were isolated using a multivariate regression model, with the outcome recorded in table 4.1.

First, in terms of issue framing, the difference between these two scales is potentially significant when they are used as dependent variables. We might, for example, expect affluent respondents to express concern for the environment in general, but due to the quality of their own surroundings believe those issues to present little direct risk to themselves. Based on theories of gender socialization and perceived vulnerability to risk, we might also expect to find differences between men and women to be stronger when environmental dangers are defined in personal terms. Those expectations, however, fail to materialize. While each set of questions might seem to tap different emotions and considerations, the scales themselves are extraordinarily correlated (with a Pearson's *r* of .91), which suggests that respondents fail to see a meaningful distinction between the impact those problems have on the environment in general and on their own personal lives.

As for the comparative importance of various social and demographic traits, age, education, gender, and partisanship all move in expected directions, and all easily reach conventional levels of what scholars call "statistical significance"—a standard often used to express the degree of confidence we have in a relationship between two variables.[45] In this sense, current results confirm the work of others who argue that environmental concern is somewhat stronger (in a relative sense) among the well educated, women, liberals, and Democrats.

Table 4.1

Explaining Variance in Perceptions of Environmental Danger

Independent variables	Model 1 General perceptions of environmental danger		Model 2 Personal perceptions of environmental danger	
	Slope coefficient	Standard error	Slope coefficient	Standard error
Age cohort:				
Born after 1971	1.00	0.71	1.18	0.77
Born 1959–1971	0.65*	0.32	1.03**	0.35
Born 1946–1958	0.90**	0.32	1.28***	0.34
Born before 1946	0.00	—	0.00	—
Education	0.15**	0.05	0.09	0.05
Income	−0.10***	0.03	−0.13***	0.03
Race				
Black	0.49	0.43	1.21**	0.47
Other	1.61*	0.66	1.75*	0.70
White	0.00	—	0.00	—
Gender				
Female	1.28***	0.26	1.26***	0.28
Male	0.00	—	0.00	—
Party identification	0.17*	0.07	0.17*	0.07
Political ideology	0.29**	0.10	0.31**	0.11
Intercept	17.69	0.78	17.72	0.83
Mean on additive scale	21.46		20.58	
Number of cases	921		922	
R-square	0.102		0.113	

Source: NORC General Social Survey (1994).

Notes: All estimates were obtained using ordinary least squares. See appendix for question wording and scale construction.

* $p < .05$.

** $p < .01$.

*** $p < .001$.

Probing further, however, reveals several surprises. While some scholars predict a linear relationship between age and environmental concern, results segmented into birth cohorts demonstrate a single ledge dividing those over the age of fifty from their younger counterparts. In other words, data confirm that respondents who came of age after 1970 have been more effectively socialized as a group into the mainstream environmental movement than have their parents and grandparents but that little added growth has occurred since that time.

Second, while income coefficients are statistically significant in both equations, its direction defies conventional wisdom in the sense that low-income respondents are more likely to see environmental problems as dangerous. While Mary Douglas and Aaron Wildavsky argue in *Risk and Culture* (1982) that environmentalism hobbles economic growth and that environmental regulations fail to serve the interests of the lower middle class,[46] empirical results here suggest those assumptions are false, perhaps because lower-middle-class families are frequently the victims of environmental pollution.[47]

Finally, while previous work has noted a racial gap on environmental attitudes, results are also somewhat surprising in suggesting that minority respondents are more responsive to environmental risks, especially when those issues hit close to home. Perhaps as Robert Bullard and others in the burgeoning field of environmental justice note, people of color (like those of low income) are more likely to experience pollution firsthand and therefore develop a heightened sensitivity to those concerns.[48]

In the end, however, the practical consequence of group differences in both equations should not be overemphasized for two reasons. Slope coefficients (while statistically significant) are, in a substantive sense, small in size relative to the distance of the scale used. To place that criticism into an appropriate context, recall that the additive scales employed in each model range in value from six to thirty points. Holding all else constant, then, a respondent who embraces a strong liberal ideology would differ from a strong conservative by less than two points across six survey items, hardly a monumental difference of opinion. Moreover, the percentage of variance in environmental concern explained by these variables in sum total remains rather poor, weighing in at just 10 percent and 11 percent respectively.[49] Ironically, in this case the failure of the model to provide a better fit reinforces the most important insight of all. Minor variations in degree notwithstanding, nearly everyone is concerned about the environment.

Environmental Policy Preferences
A second method of gauging environmental concern often involves asking respondents in surveys whether they support increased government spending and taxation for environmental programs. Like above, data for this test were drawn from the 1994 NORC General Social Survey (GSS), where respondents were questioned as to whether they felt that spending on "improving and protecting the environment" was currently "too little," "about the right amount," or "too much." They were likewise asked to gauge their willingness-to-pay "much higher taxes" for environmental purposes. Each of those measures was independently regressed on the same set of political and demographic characteristics used above, this time for statistical reasons using a technique known as *ordered probit*.[50]

While we might expect cleavages between social groups to be more pronounced here given an increased demand for commitment and sacrifice, results are much the same. Once again, younger birth cohorts are somewhat more likely to support policy efforts, and education and income have expected effects on taxation. As table 4.2 demonstrates, however, cleavages based on party identification and political ideology dominate environmental policy preferences. As Klineberg and his colleagues note, the strength of those relationships is hardly surprising and likely reflects the way in which these issues are framed in the GSS questionnaire. By relating environmental protection to government spending and taxation, environmental concern becomes cognitively linked to political ideology and long-standing beliefs about the proper role of the government sector. In fact, "the consistent differences between liberals and conservatives on these items," they write, "may have at least as much to do with the reactions to increased government intervention in general as with differences in their concerns about environmental issues per se."[51]

Table 4.2
Explaining Variance in Environmental Policy Preferences

Independent variables	Model 1 Support for increased environmental spending		Model 2 Willingness to pay higher taxes	
	Probit coefficient	Standard error	Probit coefficient	Standard error
Age cohort:				
Born after 1971	0.46*	0.20	0.47*	0.18
Born 1959–1971	0.64***	0.09	−0.01	0.08
Born 1946–1958	0.55***	0.09	−0.08	0.08
Born before 1946	0.00	—	0.00	—
Education	0.01	0.01	0.05***	0.01
Income	−0.01	0.01	0.01*	0.01
Race				
Black	0.07	0.12	−0.23*	0.10
Other	0.08	0.19	0.11	0.16
White	0.00	—	0.00	—
Gender				
Female	0.03	0.07	0.02	0.06
Male	0.00	—	0.00	—
Party identification	0.11***	0.02	0.06**	0.02
Political ideology	0.08**	0.03	0.10***	0.03
Intercept 1	−0.82	0.21	−3.15	0.21
Intercept 2	1.19	0.05	1.26	0.06
Intercept 3	—	—	1.85	0.07
Intercept 4	—	—	2.71	0.08
Number of cases	1,254		1,141	
Log-likelihood	−1010.70		−1666.17	

Source: NORC General Social Survey (1994).
Note: See appendix for question wording.
*$p < .05$.
**$p < .01$.
***$p < .001$.

Strength of Feeling toward Environmentalists

A third way of operationalizing environmental concern focuses on the feelings respondents have towards environmentalists, as opposed to the issues they support or the government policies designed to address those problems. Scholars have long argued that group identification is vital in understanding the development of social identity.[52] Indeed, in *The American Voter* (1960), Angus Campbell and his team point out that psychological attachments to groups "can become reference points for the formation of attitudes and decisions about behavior."[53]

To explore whether those considerations matter, data were selected from the 1996 National Election Study (NES) where respondents were asked to react to the term *environmentalists* by reference to a "feeling thermometer." Tapping affective (as opposed to cognitive) evaluations, movement up (positive) and down (negative) the scale was used to express feelings of warmth or coolness toward the object in question.

Not surprisingly, environmentalists fared well on the thermometer scale overall, rating an average score of 63, well above that obtained for labor unions, big business, or the federal government in Washington— all typical competitors in the arena of environmental politics. More significant than magnitude or rank, however, is the fact that no significant differences in opinion emerge between social and demographic groups listed in table 4.3, except for two—party identification and political ideology.

To simplify that result, differences in means between party identifiers are highlighted graphically in figure 4.1 in comparison to the overall distribution of responses received. While Klineberg, et al. suspect that political variables have less impact on questions that look beyond issues of government intervention, the continued strength of those variables in this context (both statistically and substantively) suggest that the term *environmentalist* retains a clear political context and connotation that may limit its appeal. In the end, those relationships likely reflect the difficulties that citizens have in separating the environment as an issue from the policies used to protect it.

Table 4.3
Explaining Variance in Feelings toward Environmentalists

Independent variables	Slope coefficient	Standard error
Age:		
Born after 1971	1.28	2.49
Born 1959–1971	0.33	1.39
Born 1946–1958	0.07	1.41
Born before 1946	0.00	—
Education	−0.08	0.38
Income	−0.00	0.11
Race:		
Black	0.72	2.12
Other	5.51	3.60
White	0.00	—
Gender:		
Female	0.00	—
Male	1.91	1.12
Party identification	1.85***	0.33
Political ideology	3.30***	0.51
Intercept	42.48	2.62
Mean on thermometer scale = 63.41		
Number of cases = 1,082		
R-square = 0.162		

Source: National Election Study (1996).
Notes: All estimates were obtained using ordinary least squares. See appendix for question wording.
* $p < .05$.
** $p < .01$.
*** $p < .001$.

Self-Reported Environmentally Responsible Behaviors

One final test of the demographic correlates of environmental concern focuses attention back on the 1994 NORC General Social Survey (GSS), which at the close of its environmental battery measured to extent to which individuals were willing to act on their environmental beliefs. Respondents were asked how often they make a "special effort" to sort glass, plastic, and paper items for recycling and how frequently they

Percent responding

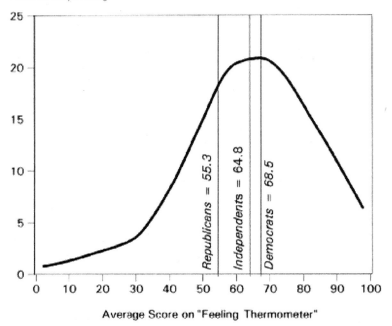

Figure 4.1
Partisan Comparisons on Feelings toward Environmentalists
Source: National Election Study (1996).
Note: See appendix for question wording.

purchase fruits and vegetables grown without the use of pesticides or chemicals. Respondents, too, were asked how often they "cut back on driving a car for environmental reasons." Using a four-point response format to grade the frequency of participation in these activities (ranging from "never" to "always"), answers were summed into an additive scale of economic behavior that was regressed once again on the same set of political and demographic characteristics used above.

In addition, the GSS questioned respondents about their current and past patterns of political activism. They were asked whether they were a member of "any group whose main aim is to preserve and protect the environment" and whether they had signed an environmental petition, given money to an environmental group, or taken part in a protest or demonstration about an environmental issue in the last five years. Given yes or no answers to each of these questions, a four-point additive scale was created and regressed one final time on an identical set of characteristics using an ordered probit model. Results for both of these equations are reported in table 4.4, and seem to warrant several conclusions.

First, although environmental concern and support for government spending are both stronger, on average, among younger age cohorts, those respondents are significantly less likely to recycle, buy organic produce, or cut back on the use of an automobile. Second, education clearly influences participation in environmentally responsible activities across the board, confirming that while respondents at all levels of education are concerned about environmental impacts close to home, those with higher degrees are better equipped to translate that concern into effective action. Third, minority respondents are somewhat more concerned about the environment than their white counterparts, but they are less likely to participate in both forms of environmentally responsible behavior, confirming previous work on barriers to participation offered by Paul Mohai and others.[54]

Fourth, despite the fact that some GSS measures (such as environmental membership and contributions to environmental groups) require an ability to pay in addition to a willingness to do so, results by and large fail to show income differences for reasons that may have much to do with the specific nature of the questions asked. For example, as noted in chapter 2, recycling demands more time than money, especially with convenient curbside pickup in many communities, and while organic produce demands a premium price tag in most supermarkets, cutting back on the use of an automobile can help to reduce energy and transportation costs, especially in areas where public transportation is readily available.

Finally, while ideology has a strong impact on political activism, those cues fail to materialize for individual economic behavior. In other words, while political conservatives seem reluctant to support liberal environmental policies, they are not unwilling to take other steps to preserve and protect the environment, especially when those activities are compatible with conservative views that promote laissez-faire or market-based approaches to environmental management.

Table 4.4

Explaining Variance in Environmental Behavior

Independent variables	Model 1 Economic behavior		Model 2 Political activism	
	OLS slope coefficient	Standard error	Ordered probit coefficient	Standard error
Age cohort:				
Born after 1971	−0.83*	0.33	0.09	0.21
Born 1959–1971	−0.86***	0.15	0.02	0.09
Born 1946–1958	−0.26	0.15	0.24**	0.09
Born before 1946	0.00	—	0.00	—
Education	0.10***	0.02	0.11***	0.01
Income	−0.01	0.01	0.01	0.01
Race:				
Black	−0.40*	0.20	−0.58***	0.13
Other	0.71*	0.30	−0.11	0.18
White	0.00	—	0.00	—
Gender:				
Female	0.28*	0.12	−0.04	0.07
Male	0.00	—	0.00	—
Party identification	0.04	0.03	0.04	0.02
Political ideology	0.10*	0.05	0.15***	0.03
Intercept 1	4.81	0.38	−5.26	0.30
Intercept 2	—	—	1.06	0.15
Intercept 3	—	—	1.88	0.16
Intercept 4	—	—	2.56	0.16
Number of cases = 999/1093	R-square = 0.074		Log-likelihood = −1090.32	

Source: NORC General Social Survey (1994).
Notes: See appendix for question wording.
*$p < .05$.
**$p < .01$.
***$p < .001$.

Conclusions

Accusations of elitism are as old as the environmental movement itself and have become a fierce weapon in the war over protective environmental policies. Some critics insist that the priorities of the environmental movement remain dangerously entrenched to the left of a "deep ideological divide" and stand at odds against prized American values like personal liberty, property rights, and technological optimism.[55] Still others, such as William Tucker in *Progress and Privilege* (1982), contend that environmentalism "favors the affluent over the poor, the haves over the have-nots" and that it "hobbles economic growth" and constrains "other people's economic opportunities."[56]

Yet as William Schneider notes, to assume that this viewpoint extends to mass politics represents a "serious miscalculation."[57] As data presented in this chapter show, demographic factors once thought to be predictive of environmentalism (such as age, education, and income) no longer characterize the distribution of environmental attitudes in the United States very well.[58] When actual disagreement among respondents is distinguished from artificial variance introduced through question wording and survey design, partisanship and ideology remain the only consistent cleavages dividing Americans in their opinions on the environment, and even when taken together those models rarely account for more than a small fraction of the variance observed.[59] To assume, then, that environmental concern represents the privilege of a "new class" of social and economic elites (as Tucker does) is to misjudge the breadth of its appeal. Because of the wide scope of environmental problems, from wilderness and wildlife preservation to air pollution and water contamination, environmentalism is valued by nearly everyone.

In the end, of course, that basic truth represents an important victory for environmentalists, one that in Robert Gottlieb's words expands the definition of environmentalism and "broadens the possibilities for social and environmental change."[60] A growing link between populism and the environmental movement not only lends credibility to the claim that environmentalists represent the public interest; it suggests that those values may have the power and potential with time to develop into a fundamentally new social paradigm or belief system.[61]

5
Constraint: Are Environmental Attitudes Inconsistent?

When published in 1964, Philip Converse's influential article on "The Nature of Belief Systems in Mass Publics" offered an extraordinarily bleak view of public opinion.[1] In it he stressed the importance of internal consistency or "constraint" among different attitudes, arguing that without a true anchoring mechanism—such as ideology—the events of political life could become a confusing array of disconnected facts.[2] But in place of the logical coherence he prized, Converse found mainly chaos. In pointing to what seemed like clear evidence in surveys, he confessed in the end that large portions of the American electorate not only failed his test, many lacked meaningful beliefs at all, "even on issues that have formed the basis for intense political controversy."[3] It was, as one team of scholars later put it, as if respondents in polls "indulge interviewers by politely choosing between the response options put in front of them—but choosing in an almost random fashion."[4]

Today, the evidence cited by Converse is considered one of the most unsettling findings in modern research on public opinion, opening the door to a firestorm of controversy.[5] But it is also a topic that underscores an important gap in our analysis of the environment so far. By focusing on characteristics such as direction, strength, stability and distribution, a wide range of survey instruments has been considered but almost always independently, one issue at time. According to V. O. Key that means that "an important aspect of public opinion is obscured," one that rightly demands equal attention be paid to cognitive relationships *between* attitudes.[6]

On that count, the stakes could not be higher. For environmentalists convinced that social values are at "the root of the ecological crisis,"[7] growing environmental concern would seem to give hope that a major transformation is at hand "in the American public's beliefs about how the world works physically, socially, economically, and politically."[8] Yet finding persuasive evidence in support of that claim is difficult. When it comes to attitude consistency, survey researchers have long puzzled— as Converse did—over low correlations between different measures of environmental concern, at times suggesting that public demands weave a veritable crazy quilt of conflicting ideas and desires.

Why should various survey instruments appear idiosyncratic? In part, the answer could be one of faulty survey design. Perhaps lenient questionnaires too readily encourage respondents to proclaim themselves environmentalists, only to back away from environmental goals when painful trade-offs and behavioral commitments are brought to mind. As data presented in this book have suggested already, Americans are sympathetic to environmental problems, but most remain unwilling to act on their beliefs either as voters, consumers, or political activists.[9] A continued gap between attitudes and behavior—between what Americans *say* and what they *do*—is consistent with the criticism that people lack a strong underlying orientation toward the environment, and that surveys that purport to measure environmental attitudes find little more than "doorstep opinions" conditioned by social desirability.[10]

Second, the reason could lie in the structure of environmental opinion itself. Based on ideology, it might be reasonable for some respondents to oppose government policies on environmental issues they nonetheless care deeply about. Others might react to environmental conditions in their own local communities by expressing concern for a singular issue, such as water pollution or toxic waste disposal, without being swayed by others that fall into the environmental rubric. In other words, attitudes toward the environment could be truly multidimensional, "splintered," as Lance deHaven-Smith says, "into a number of separate and narrowly focused belief systems," where individual issues are dealt with in terms of unique symbols and reference points.[11]

One final possibility remains, consistent with several themes developed so far. When observing cognitive inconsistency, is attitude or methodology to blame? Scholars have long recognized that survey instruments measure opinions imperfectly for a variety of reasons that are both mundane and accidental, creating a problem known as *measurement error.* Mathematics demonstrate that when those errors become systematic, the correlations observed between survey items can be misleading, even inaccurate. In the end, all are possibilities that are explored (and compensated) in detail below.[12]

The Dimensionality Problem Defined

The first generation of articles on environmental attitudes began to appear in academic journals soon after the first Earth Day in 1970. Because the field was new and because survey researchers were developing questionnaires independently and administering them locally, they tended to word questions in ways that were very different. While some studies measured attitudes toward specific environmental problems, such as air pollution or wildlife protection,[13] others choose to examine support for government spending,[14] knowledge about environmental issues,[15] preferences for environmental policy,[16] or self-reported participation in environmentally responsible activities, like recycling or energy conservation.[17] With few exceptions, most studies treated issues on that diverse list as indicators of the same underlying trait, something broadly termed *environmental concern*.[18] Since that time, however, a second generation of articles in the field has noted (and struggled with) low correlations between measures, drawing considerable attention to issues of measurement and finally to the question of dimensionality itself. It is a debate that strikes to the heart of Kent Van Liere and Riley Dunlap's question: "Does it make a difference how environmental concern is measured?"[19]

For most scholars, the answer has been a resounding "yes."[20] Many argue that environmental attitudes are issue-specific, and that the same thoughts and ideas are not being tapped in each case.[21] One research team found a strong relationship among five attitudinal scales (with correlation coefficients ranging from .53 to .81),[22] but a majority of studies suggest instead that environmental items factor into two or more unique dimensions, at times supported by different cultural and socioeconomic groups.[23]

Broader studies attempting to quantify a paradigm shift that is more sensitive to the needs of the natural environment have reached a similar impasse. Van Liere and Dunlap chart a major shift in social values and beliefs away from what they term the "dominant social paradigm"

(DSP)—one centered around abundance and progress, devotion to growth and prosperity, faith in science and technology, and commitment to a laissez-faire economy.[24] They believe that public attention to potentially catastrophic trends has undermined the assumptions of the DSP, pushing social norms in the direction of a "new environmental paradigm" (NEP) that comes to grips with limits to growth, the balance of nature, and the finite availability of natural resources.[25] Using survey data from Washington state, they demonstrate that all twelve items in their battery of questions cluster strongly together into a coherent and logical belief system.[26]

Unfortunately, several attempts to replicate those results have failed. In using identical measures on two additional Iowa samples, one team concludes that the NEP items are not singular but rather break down into three distinct dimensions they identify as "balance of nature," "limits to growth," and "man over nature."[27] In using a more sophisticated model, Jack Geller and Paul Lasley confirm neither outcome, arguing instead in favor of a three-factor, nine-item model truncated from the original.[28] Still other attempts have asserted that the scale breaks down into two, three, or even four dimensions.[29] It is a problem, in part, that recently prompted Dunlap and his colleagues to develop a new and "improved measuring instrument."[30]

Given countless studies such as these that underscore the inconsistency of environmental attitudes, some students of mass belief systems have gone one step further by viewing multidimensionality as lack of constraint. Much as Converse did, they suggest that public attitudes on the environment are not rooted in abstract philosophical or ideological principles but are instead rather crude and disconnected, "splintered into a plethora of narrowly focused perspectives."[31] If true, this conclusion challenges not only the validity of many of the environmental measures used in surveys but also the very existence and utility of a concept broadly termed *environmentalism* in American mass politics.

Methodological Considerations

While previous research has highlighted the incongruous nature of public attitudes on the environment, the methodology of some of this literature is problematic for several reasons. With the exception of studies on the new environmental paradigm (NEP) scale, most researchers have used independent question wording that hinders direct comparison. The samples used in many surveys also have been small, regionally based, and self-administered.[32] The data explored in this chapter offer improvement in both respects. Not only does the Gallup Organization provide a well-drawn national sample, avoiding the self-selection bias found in most mail surveys, but each of the questions used was repeated using identical wording at least six times since 1989, allowing for replication to test the robustness of results.

Second, most work on the dimensionality of environmental concern overlooks the importance of *measurement error*—a term that refers to a number of related and potentially serious problems. In some cases, pollsters may unwittingly commit errors in coding; in others, respondents may interpret the questions posed to them in idiosyncratic ways or find that they are unable to communicate their views accurately given the crudeness of the response categories presented.[33] Errors in measurement that occur randomly, such as these, tend to attenuate correlations between survey instruments. In other words, it generates correlation coefficients that are weaker than they should be.

In most previous studies on environmental attitudes, however, a far more pressing problem looms since batteries of questions tend to be used in close proximity using an identical response format.[34] Given that the same approach is used here for lack of a better alternative, some explanation (and correction) is required. Recall, for example, that in the Gallup study (March 5–7, 2000) introduced in chapter 2, respondents were asked if they personally worried "a great deal," "a fair amount," "only a little," or "not at all" about each of a number of environmental problems. While the subject changes in each variation of the question, the appropriate set of responses remains static. An accumulation of experience in the field of survey research has shown that under those conditions respondents have a tendency to anchor themselves along the response continuum, answering each subsequent question in that battery relative to some personal reference point.[35] The ensuing problem—formally described as error covariance—is an important one since mathematics demonstrates that results contaminated by nonrandom measurement errors are unpredictable. The correlations calculated between different survey questions can be either larger *or* smaller than the true correlation, and may even be of the wrong sign.[36]

In short, when the likelihood of measurement error is introduced into the issue of attitude constraint on the environment, two oddly opposing possibilities arise. If correlations between survey questions are due to their close proximity and common response format alone, a strong relationship may shatter when those errors are purged from the data. On the other hand, it is plausible for measurement error to have an opposite effect. It might cause attitudes to appear multidimensional when in reality the relationship between different environmental opinions is highly constrained. Since the possibility of nonrandom measurement error cannot be accommodated by looking at correlation coefficients alone, a technique known as *confirmatory factor analysis* is used here instead. Flexible enough to allow different sources of error to be estimated and controlled, it is an ideal approach for this situation.[37]

In general, the term *factor analysis* refers to a family of data reduction techniques designed to remove redundancy from a set of interrelated variables by clustering them together into common elements, called *factors*.[38] It tells researchers, in part, "what tests or measures belong together"—that is, "which ones virtually measure the same thing."[39] The approach adopted in this chapter can be described as confirmatory because it allows for the development of empirical models that test and compare alternative hypotheses using standard goodness-of-fit statistics. Using prior knowledge and theoretical expectation as a guide, the goal is to determine the number of factors that underlie a given set of survey instruments—one, two, three, or more in the case of the NEP scale—and to record the strength with which those elements are related to one another.

Data Analysis

In a practical sense, the issue of dimensionality can be broken down into at least two component parts. The first addresses the consistency of responses across different environmental problems. If a respondent is concerned about air pollution, for example, are they likely to feel concern for other environmental issues as well, such as water pollution, acid rain, and global warming? A second question regarding the dimensionality of environmental attitudes is wider in scope and potentially more significant, highlighting the relationship between different idea-elements thought to define the concept of environmentalism. Those elements might include the degree to which Americans worry about the seriousness of environmental problems, their preferences on environmental policy, and their willingness to identify themselves as active or sympathetic members of the environmental community.

To examine the consistency of environmental attitudes across measures such as these, data for this chapter were drawn from a March 2001 study administered by the Gallup Organization. Recall once again that within a lengthy battery of environmental questions, respondents were asked whether they personally worry "a great deal," "a fair amount," "only a little," or "not at all" about thirteen different environmental issues, ranging from air pollution to the deforestation of tropical rain forests. In addition, respondents were asked to express their support for (or opposition to) eight specific environmental policy proposals, including "expanding the use of nuclear energy," "opening up the Alaskan Arctic Wildlife Refuge for oil exploration," and "spending more government money on developing solar and wind power." Finally, survey participants were asked by Gallup if they thought of themselves as "an active participant in the environmental movement" or as merely "sympathetic toward the environmental movement, but not active." The close-ended response format for this question also included a position of neutrality and one that was admittedly "unsympathetic" to the environment.[40]

Selection of Variables

It should be noted from the outset that some variables included in the full Gallup battery were discounted from analysis here for several reasons.[41] The first decision was essentially a pragmatic one. A smaller subset of variables made model identification and convergence more manageable and ultimately the results more parsimonious. Second, to test the consistency of environmental beliefs, two strong and competing models had to be developed to group clusters of variables based on instinct and expectation. Some variables fit that purpose more naturally than others. For instance, attitudes toward air and water pollution, along with acid rain, most likely contain common elements whereby all three indicate (and jointly measure) concern for environmental pollution. Issues such as ozone depletion, deforestation, and the "greenhouse effect" would seem to represent a broader concern for global environmental problems. Some issues, such as urban sprawl, were excluded from both models because they seemed to bring to mind considerations that were substantially different.

Third, given the importance of test statistics in discriminating between competing models, variables with degrees of non-normality severe enough to cause problems were excluded from consideration also. Confirmatory factor analysis requires a number of strict distributional assumptions. Non-normality in the form of excessive skewness or kurtosis can threaten the validity of significance tests and goodness-of-fit statistics, such as chi-square.[42]

Finally, variables were chosen with replication in mind. Gallup's battery of environmental concern items has changed slightly in its many permutations since 1989, mainly through the addition of new items. To allow direct comparison with previous work (which enhances the robustness of results), only variables that are repeated in all versions of the study since 1989 are included here.

A Comparison of Two Models

With the above decision rules in place, the following analysis concentrates initially on a subset of six environmental problems. Descriptive statistics for these variables appear in table 5.1. To gauge the strength of the relationship between each set of issues, two models are compared. In the first, environmental concern is viewed as "multidimensional" and is represented by two factors. One uses three survey items to target concern for environmental pollution, while the other uses an additional three to weigh global environmental problems. An alternative "unidimensional" model posits that all six measures are indicators of the same underlying trait—that is, that all six questions essentially measure the thing. Because all of the issues in Gallup's degree-of-concern battery employ the same response format, measurement error also must be incorporated into each

Table 5.1
Correlation Matrix for Thirteen Measures of Environmental Concern

I am going to read you a list of environmental problems. As I read each one, please tell me if you personally worry about this problem a great deal [4], a fair amount [3], only a little [2], or not at all [1].

		Mean	Correlation with an additive scale of all items
V1	Ocean and beach pollution	3.15	0.74
V2	Air pollution	3.23	0.77
V3	Acid rain	2.70	0.79
V4	Damage to the earth's ozone layer	3.10	0.84
V5	The loss of tropical rain forests	3.13	0.75
V6	The "greenhouse" effect or global warming	2.86	0.83

Item Intercorrelations

	V1	V2	V3	V4	V5	V6
V1	1.00					
V2	0.56	1.00				
V3	0.49	0.54	1.00			
V4	0.52	0.60	0.58	1.00		
V5	0.51	0.47	0.49	0.54	1.00	
V6	0.49	0.54	0.60	0.72	0.53	1.00

Source: Gallup Organization (March 5–7, 2001).
Note: Given a listwise deletion of missing values used throughout, the effective sample size is 1,019 adults nationwide.

model. By estimating error variances for each variable in addition to a single error covariance term, correlations between factors can be effectively purged of random mistakes and systematic biases.

Empirical results for both models are displayed visually (and for sake of a general audience in a somewhat simplified form) in figure 5.1. Two conclusions merit special attention. First, reducing the data down to a single factor is largely successful. Results based on confirmatory factor analysis can be evaluated and compared based on the overall fit of each model using a variety of statistics, such as the ratio of chi-square to degrees of freedom.[43] That test reveals an acceptable fit for both models, as well as equal and impressive goodness-of-fit indices,[44] and sufficient

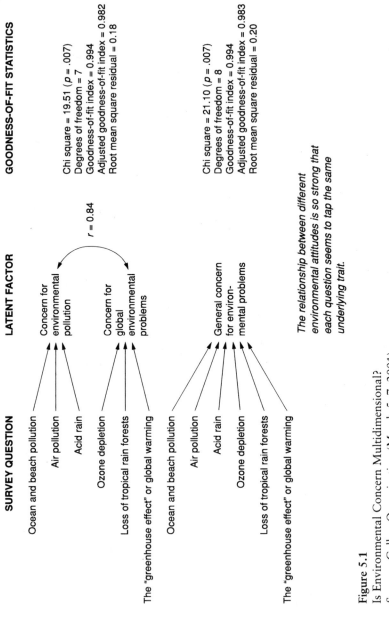

Figure 5.1
Is Environmental Concern Multidimensional?
Source: Gallup Organization (March 5–7, 2001).
Note: n = 1,019 adults nationwide. A listwise deletion of missing values is used throughout. Since survey questions contain some degree of measurement error, error variances for each variable were estimated, and a single error covariance was designed to purge the data of response set bias. See table 2.1 for question wording.

root mean square residuals, all of which suggest that the multidimensional version fails to offer new information or significant improvement.[45]

Finally, and most important of all, the estimated correlation between the two traits of interest—concern for environmental pollution and concern for global environmental problems—is an impressive .84. Indeed, that result is persuasive not merely because of its objective strength but because of its relative improvement over initial correlations between survey questions (where correlation coefficients averaged just .55). That difference in strength can apparently be attributed to the damaging consequences of measurement error. For example, if no errors in measurement are assumed—that is, if we assume that environmental attitudes are communicated perfectly by respondents—the correlation between traits using the same model is .65. If random errors alone are assumed, the same correlation rises to .77. In this case, given the added effect of a common response set, modest relationships between variables are transformed, highlighting the extent to which Americans possess a single unifying orientation toward the environment.[46]

A Broader Model of Environmental Attitudes
In addition to questioning respondents about their personal concern for a variety of environmental problems, the March 2001 Gallup study also included a series of measures on environmental policy preferences, as well as a single question probing the willingness of respondents to identify themselves with the environmental movement. As a final step in testing the dimensionality of public attitudes on the environment, the relationship between these broader idea-elements is examined below using the same strategies and techniques.[47]

As results for this model demonstrate in figure 5.2, correlations between these three factors are also quite high, especially relative to those produced under unreasonable measurement assumptions. Concern for environmental issues correlate with environmental policy preferences at .67. Correlations between environmental self-identification and concern and between self-identification and policy preferences measure .51 and .49, respectively. In this case, the presence of both random and nonrandom measurement error produced correlation coefficients that were just 60 to 70 percent of their true value.[48] Clearly, as with the two

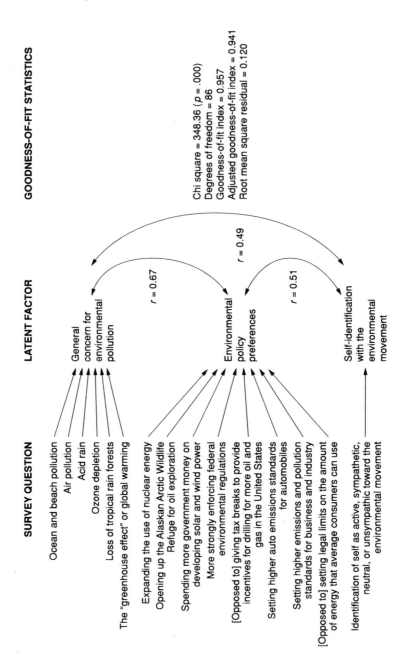

SURVEY QUESTION

LATENT FACTOR

GOODNESS-OF-FIT STATISTICS

Ocean and beach pollution
Air pollution
Acid rain
Ozone depletion
Loss of tropical rain forests
The "greenhouse effect" or global warming

General concern for environmental pollution

Expanding the use of nuclear energy
Opening up the Alaskan Arctic Wildlife Refuge for oil exploration
Spending more government money on developing solar and wind power
More strongly enforcing federal environmental regulations
[Opposed to] giving tax breaks to provide incentives for drilling for more oil and gas in the United States
Setting higher auto emissions standards for automobiles
Setting higher emissions and pollution standards for business and industry
[Opposed to] setting legal limits on the amount of energy that average consumers can use

Environmental policy preferences

Identification of self as active, sympathetic, neutral, or unsympathic toward the environmental movement

Self-identification with the environmental movement

r = 0.67

r = 0.49

r = 0.51

Chi square = 348.36 (p = .000)
Degrees of freedom = 86
Goodness-of-fit index = 0.957
Adjusted goodness-of-fit index = 0.941
Root mean square residual = 0.120

models reported earlier, "corrected" correlations obtained using confirmatory factor analysis result in a substantially different (and more hopeful) view of the consistency of environmental attitudes across multiple measures.

Conclusions

Students of environmental opinion have long tried to understand why different measures of environmental attitudes fail to correlate more strongly, going so far as to argue that assumptions about unidimensionality are both "unwarranted and misleading."[49] While that issue might be considered little more than a narrow methodological curiosity, it is work that in recent years has gained increased importance by intersecting with studies on mass environmental belief systems, with the two together suggesting that public attitudes on the environment are rather crude, disconnected, and narrowly focused.[50]

Yet the dimensionality problem may, at least in part, be an artifact of faulty methodology. As data presented in this chapter demonstrate, various measures of environmental concern are not as unrelated as previous studies suspect. By using confirmatory factor analysis to control for both random and nonrandom sources of measurement error, not only can the environmental battery used by Gallup be reduced to relatively few dimensions; those dimensions are themselves strongly correlated.

When studying environmental attitudes, then, does it truly "make a difference how it's measured"?[51] Of course, the answer is yes. Despite strong correlations between factors, current results underscore the im-

◀ Figure 5.2
Are Different Types of Environmental Attitudes Related?
Source: Gallup Organization (March 5–7, 2001).
Notes: $n = 1,019$ adults nationwide. A listwise deletion of missing values is used throughout. Since survey questions contain some degree of measurement error, error variances for each variable were estimated, in addition to a single error covariance term for each battery designed to purge the data of response set bias. To achieve model identification, the error variance for Gallup's measure of self-identification was fixed so that its reliability was equivalent to its Cronbach's coefficient alpha.

portance of measurement assumptions in survey design and data analysis. Consistent with a growing body of research that has demonstrated the importance of controlling for systematic response biases when evaluating political and psychological attitudes,[52] data results here demonstrate how errors in measurement can effectively disguise strong correlations between various environmental measures that might otherwise appear modest or inconsistent.

More important, however, empirical results in this chapter raise important questions about the existence and sophistication of mass environmental belief systems. If, as Lance deHaven-Smith suggests, multidimensionality should be viewed as lack of constraint, evidence of attitude stability across multiple measures seems to offer compelling evidence that public attitudes on the environment have matured into a logical, structured, and constrained belief system. Considerable care, however, should be taken in drawing conclusions about the quality and sophistication of those beliefs.[53] Given that most attitudes on complex environmental issues can be represented by relatively few dimensions, the models developed here may signify only that it is "cognitively economical" for people to reduce those concerns into a general environmental orientation, regardless of knowledge or clear reasoning.[54] In other words, as Christopher Achen warns, a "certain stability of viewpoint is a necessary, but hardly sufficient condition for political understanding."[55] In the end, consistent attitudes on the environment may suggest that Americans are increasingly willing to express concern for environmental quality, but dimensionality alone may say little about their readiness to become active, well-informed participants in the environmental movement.

II

Behavior

6

The Ballot Box I: Issue Voting and the Environment in Presidential Elections

In 1980, in response to a long-standing debate about the role that issues play in presidential campaigns, Edward Carmines and James Stimson described what they called "two faces" of issue voting.[1] On the one hand, they said, some topics on the political agenda were especially "hard" for voters to comprehend, in particular those that were little known, or that demanded sophisticated calculation or technical expertise. On the other were issues that were "easy" in comparison, those to which reactions were naturally symbolic and emotional; those that invited a "gut response" to long familiar issues, where the ends of public policy were in question rather than its means—in short, issues very much compatible with popular thinking about the environment.

Since Carmines and Stimson believed that "easy" issues impose fewer cognitive demands on voters, their typology suggests that voting based on the environment is possible, even likely. But to the contrary, scholars have long noted the near absence of those issues in national political campaigns.[2] In fact, environmental failures at the ballot box seemed so transparent in the 1980s and 1990s that scholars and political pundits alike seemed ready to dismiss the subject as a political "paper tiger" that was long on talk but short on action.[3] As one critic quipped,

Forget the hundreds of polls showing that 80 percent of Americans would walk over their grandmothers' graves to save a tree. No other lobby has been as routinely and overwhelmingly rejected at the polls during the past six years as has the environmental movement. When will Washington realize that the Green Emperor has no clothes?[4]

Expectation and conventional wisdom aside, however, there has been remarkably little systematic study of environmental preferences as an influence on electoral choice. With only limited survey data available, the few studies to address environmental voting in the United States do more to report a deficiency than to explain why it should be the case.[5] Yet it is a topic rich in opportunity. With a Republican-led legislature since 1994 scaling back wildlife protection and pollution control laws, an increasing number of poll watchers in the popular press have predicted that the environment will emerge at long last as a potent political weapon for Democrats, in particular one that can be used as a "wedge" issue to attract young, socially moderate voters away from the Republican Party.[6] Using a new series of variables introduced in the 1996 and 2000 National Election Studies, this chapter tests the potential of that claim by measuring the electoral impact of environmental issues on American political parties and their candidates.

Theory and Background

As Carmines and Stimson aptly point out, the study of issue voting has been "infused with normative considerations" from the start.[7] Voters who cast ballots based on their own personal policy preferences relative to those of party candidates are often assumed to make decisions that are rational, wise and sophisticated.[8] Likewise, issue voting would seem to ensure an active link between the views of citizens and those of elected officials in a way that ultimately enhances popular sovereignty and collective responsibility. By nearly every account, however, environmental issues fail to provide that link in satisfactory fashion. "So far," as Philip Shabecoff writes,

environmentalism has had remarkably little impact on electoral politics, particularly at the national level. Although people might care a great deal about the environment, they have not, at least in the past, voted for candidates largely because of environmental records or positions.[9]

The consequences of that deficiency, at least on the surface, seem clear. Without an electoral disincentive, scholars argue that congressional roll call votes on environmental policies are virtually unaffected by constituents' environmental demands.[10] Votes on environmental issues in Congress tend to split along a strong partisan divide, despite a growing consensus in the mass public that cuts across party lines.[11] Even more troubling is evidence that shows that policy divergence on environmental issues between Democrats and Republicans has grown wider still in recent years in both the U.S. House of Representatives and the U.S. Senate.[12] It is a trend, according to some, that suggests a troublesome "gap between the policy preferences of the electorate and the actions of elected representatives."[13]

In the past, it was nearly impossible to study issue voting and the environment among average Americans because of a paucity of data. That problem is rectified in this chapter using a new battery of questions developed for the 1996 National Election Study. But certain theoretical issues must be broached first. For example, why might environmental issues falter in national campaigns despite honest and enduring public concern? At least three reinforcing (and times conflicting) possibilities are explored here—issue salience, perception of candidate differences, and partisan loyalty.

Issue Salience
First, perhaps environmental issues fail to influence American voters because their preferences lack a requisite degree of intensity or personal importance. John Zaller notes this very possibility in arguing that the weak impact of the environment on candidate evaluations in the 1991 NES Pilot Study might have been an "artifact" of low salience during a year understandably dominated by foreign policy concerns during the Persian Gulf War.[14] If true, Zaller's conclusion suggests that environmental issues might generate a stronger political punch if and when Americans become convinced that the nation's environment is in crisis. Recent articles in the popular press that report on the public's growing dissatisfaction with the environmental priorities of Republican lawmakers in Congress follow this same logic.[15]

Empirical evidence on issue salience among scholars, however, is decidedly mixed. Howard Schuman and Stanley Presser argue that voters who consider an issue to be "important" are more likely to translate their convictions into political action.[16] Yet others find that "salience plays a substantial but not overwhelming role in determining candidate evaluations" and that it "cannot be deemed the sole or even the dominant factor" in understanding electoral choice.[17] Other factors must be considered as well.

Perception of Candidate Differences

A second possible explanation for why environmental issues falter in national campaigns recognizes that the likelihood of an issue vote depends on the ability of citizens to distinguish between the policy positions of the candidates.[18] In *The American Voter* (1960), Angus Campbell and his colleagues argue that for issue positions to influence individual vote choice, several cumulative conditions must be met.[19] The first condition is largely cognitive: the voter must be aware of the existence of an issue like the environment and must have formed an opinion about it. Not surprisingly, some minimal intensity of feeling (or salience) defines the second condition. Equally important, however, is the third— that is, the voter's ability to discriminate accurately between the policy positions of the two parties or their candidates. Without the latter they write, the issue can have "no meaningful bearing on partisan choice."[20]

Given a low degree of voter interest and even lower levels of political knowledge and sophistication, Campbell, Converse, Miller, and Stokes find that most voters fail to perceive party differences, even on important matters of public policy. Still, a number of scholars have contradicted that basic finding, arguing instead that issues increase in power when candidates actively articulate their policy alternatives, as during the Vietnam War.[21] In other words, clarity regarding political issues seems to depend on clarity of choice. If voters perceive few differences between candidates on matters of environmental policy, they may be forced to decide based on other issues or considerations.[22]

Partisan Loyalty

A third and final reason why environmental issues fail to impact voting behavior may be the elemental power of partisanship and its ability to condition which candidate voters see as most capable of handling environmental problems. For example, some find that judgments of party competence—central to the logic of issue voting—change slowly in response to new information and are clearly constrained by prior beliefs and long-standing partisan commitments.[23] That logic has a clear implication here. Since voting "green" often demands that loyal Republicans cross party lines to vote for liberal political candidates or strict regulatory policies, some voters may be reluctant to make those decisions on ideological grounds.

Understanding Vote Choice

For the first time in 1996, the National Election Study (NES) devoted an extensive battery of questions to environmental issues. The goal, as one pair of scholars put it, was to "embed the study of the environment in the broader context of national politics and to unpack the political consequences of the environment on the ways that citizens evaluate candidates and make vote choices in national elections."[24] By including measures that tap perceptions of environmental quality, the placement of candidates along issue scales, the ability of parties and their candidates to handle environmental problems, and the general importance of environmental issues to voters, the data set is ideally suited to the issues raised here.

The NES is uniquely valuable for one additional reason as well. With parallel instrumentation across many measures, the study allows for direct comparisons to be made between environmental issues and other social, economic, and political concerns, including poverty, health care, abortion, national defense, and so on. The extent to which environmental attitudes are similar to (or different from) opinions on other issues may help us to understand its electoral potential.

A Democratic Strength

First and foremost, data drawn from the 1996 National Election Study are unambiguous on one point. By all accounts and measures, the environment is a strong issue for the Democratic Party and its candidates. As table 6.1 indicates, although respondents (understandably) have some tendency to see their own party as best able to handle the "problem of pollution and the environment," a significant number of Republicans—35 percent of weak identifiers and 27 percent of strong—believe that the Democratic Party would do a "better job" in that area nevertheless.

Table 6.1

Party Performance on Pollution and the Environment

Which do you think would do a better job of handling the problem of pollution and the environment—the Democrats, the Republicans, or wouldn't there be any difference between them?

Partisan Identification	Democratic Party	No Difference	Republican Party
Strong Democrat	71.8%	24.7%	3.5%
Weak Democrat	48.3	46.6	5.1
Independent Democrat	51.4	43.9	4.7
Independent	29.1	57.0	13.9
Independent Republican	28.1	53.7	18.3
Weak Republican	35.0	47.0	18.0
Strong Republican	27.0	40.5	32.4

Source: National Election Study (1996).
Notes: Number of cases = 842
Chi-square = 134.5 ($p = 0.001$)
Degrees of freedom = 12
Gamma = 0.400

Numbers of nearly identical strength appear when Bill Clinton is considered as a presidential candidate, independent of his party affiliation.

Interestingly enough, data that follow in table 6.2 demonstrate that among Republicans alone, the perceived strength of the Democratic Party on environmental policy outranks *all* other issues used on the NES questionnaire. Respondents were asked to rate party competence on a range of social and economic problems, including poverty, health care, welfare, crime, foreign affairs, and the budget deficit. While few Republican identifiers placed greater relative faith with the Democrats in "handling the nation's economy" or "keeping out of war," 31 percent sided with the Democratic Party on the environment. Both of these comparative factors—across issues and among partisan groups—suggest that the latent potential for vote defection on environmental issues might well exceed other policy arenas. As David Mastio writes, it is a weakness that should be troubling to the Republican party:

Whether public opinion is out of touch with reality or not, the political fact is that to win national elections, Republicans need the votes of millions of people

Table 6.2
Democratic Party Strengths among Republican Voters

	Republicans who believe the Democratic Party would do a better job
Handling the problem of pollution and the environment	31.1%
Handling the problem of poverty	23.5
Making health care more affordable	20.1
Reforming the welfare system	9.5
Handling foreign affairs	7.0
Handling the budget deficit	6.5
Dealing with the crime problem	6.1
Handling the nation's economy	4.3
Keeping out of war	3.9

Source: National Election Study (1996).
Note: All questions were asked of random half samples, leading to relatively small sample sizes ranging from 228 to 231 respondents. All Republicans self-identified as such in v960420. No Independent "leaners" were included.

whose stated commitment to environmental protection is fully in line with such far-left groups as the Natural Resources Defense Council and Greenpeace. In no other arena of public debate have Republicans and conservatives let themselves be so thoroughly whipped for so long without stopping to figure out what went wrong.[25]

The Democratic advantage on the environment, however, is not as strong as it initially appears. An entirely different picture emerges when environmental issues are added into two regression models that statistically weigh the influence of the environment alongside other issues that compete for scarce energy and attention. The dependent variable used in each equation is simple but powerful. In 1996, the NES questionnaire asked respondents to evaluate each of the presidential candidates by reference to a "feeling thermometer." Ranging in value from 0 to 100, the scale has been used consistently over the past thirty years to measure the degree to which respondents feel "warmly" or "coolly" toward political groups and personalities—in this case, toward President Bill Clinton and his Republican challenger, Bob Dole. By regressing each thermometer scale on a series of variables in table 6.3, results show that environmental

Table 6.3
Factors Influencing the Evaluation of 1996 Presidential Candidates

Independent Variables	Bill Clinton		Bob Dole	
Political variables:				
Partisan identification	−6.53***	(0.42)	4.06***	(0.39)
Political ideology	−2.57***	(0.66)	2.79***	(0.63)
Issue positions:				
Environment/economy	0.24	(0.49)	0.21	(0.46)
Government health insurance	−0.98*	(0.44)	0.77	(0.42)
Guaranteed job/standard of living	−0.51	(0.52)	0.10	(0.49)
Services/spending	−2.56***	(0.54)	0.96	(0.50)
Aid to blacks	0.03	(0.52)	−0.91	(0.49)
Reduce crime	−0.35	(0.39)	−0.12	(0.37)
Women's rights	−0.55	(0.45)	0.45	(0.43)
Defense spending	−0.40	(0.54)	1.18*	(0.51)
Abortion rights	1.06	(0.58)	−1.09	(0.55)
State of the nation's economy	−3.95***	(0.49)	1.26**	(0.46)
Demographic characteristics:				
Age	0.00	(0.04)	0.11**	(0.04)
Education	−0.83	(0.47)	0.37	(0.44)
Income	−0.20	(0.13)	0.31*	(0.12)
Gender	−1.03	(1.33)	−0.24	(1.25)
Race	4.33	(2.35)	4.60*	(2.22)
Intercept	119.92	(5.32)	8.84	(5.03)
Mean "feeling thermometer" score	59.3		51.8	
Number of cases	891		888	
R-square	.609		.409	

Source: National Election Study (1996).
Note: All estimates were obtained using ordinary least squares (OLS) regression. Standard errors appear in parentheses. See appendix for question wording.
* $p < .05$.
** $p < .01$.
*** $p < .001$.

policy preferences all but disappear among a sea of competing issues and influences, including controls for partisan identification and political ideology. In short, despite more stable foreign policies and a strong national economy in 1996, both models largely confirm Zaller's belief that environmental issues "carry relatively little political weight, in that they add little or nothing to our ability to explain" political attitudes and outcomes.[26] The remaining analysis in this chapter is devoted to understanding why that should be the case, focusing on the three theories outlined earlier.

Why Americans Fail to Vote "Green"
Recall that a substantial body of literature in the field of political psychology posits issue voting as the end result of a cumulative process of conditions that citizens either succeed (or fail) to meet.[27] If that is the case, perhaps environmental issues do not influence electoral behavior because comparatively few respondents see differences between the candidates' positions on the environment or because those concerns fail to matter to them personally with enough intensity to override longstanding partisan commitments.

The first suspicion, at least, is easily confirmed. NES data show that while the average voter in 1996 had more in common with President Clinton on the environment than his Republican challenger, most saw relatively small differences between the two major-party candidates. Respondents were asked to consider a seven-point issue scale, where a value of one represented the belief that "it is important to protect the environment even if it costs jobs or otherwise reduces our standard of living." A score of seven selected at the opposite end of the continuum meant that "protecting the environment is not as important" as economic interests, with numbers falling in between representing a variety of compromise positions. A second, similar question was used to rate commitment to environmental regulations that place a burden on business, while several others were used to gauge policy preferences on a range of non-environmental issues, including government health insurance, crime, and defense spending. On each of these topics respondents ultimately were asked not only to place themselves along a seven-point scale but to place both of the major-party presidential candidates as well. The

Table 6.4
Average Perceptions of the Candidates' Issue Positions, 1996

Issue	Self	Bill Clinton	Bob Dole	Distance between candidates
Environment/economy	3.53	3.47	4.55	1.08
Environmental regulation	3.42	3.24	4.57	1.33
Government health insurance	3.97	2.86	5.08	2.22
Guaranteed job/standard of living	4.46	3.27	5.09	1.82
Services/spending	4.11	3.09	4.86	1.77
Aid to blacks	4.82	3.32	5.00	1.68
Reduce crime	4.46	3.70	5.10	1.40
Women's rights	2.25	2.18	3.38	1.20
Defense spending	4.02	3.95	4.65	0.70

Source: American National Election Study (1996).
Note: For question wording on seven-point issue scales, refer to the appendix.

combination of those three items creates a perception of distance, not just between candidate and respondent but between the candidates themselves—something Carmines and Stimson call a "spatial map."[28]

As table 6.4 demonstrates, at least in comparison to other issues, those responding to both environmental questions placed the candidates within minor distance from one another, with President Clinton (on average) slightly left of center, and Senator Dole at a nearly equal distance to the right. Only on the topics of women's rights and defense spending were the candidates' positions any less distinct. Indeed, a significant number of those polled were "not very certain" at all of where Clinton (30 percent) or Dole (40 percent) stood on the subject. If issue voting depends on clarity of choice, the environment would seem to be at a sure disadvantage.[29]

The low salience of environmental issues seems likely to create an obstacle to voting as well. When respondents were asked whether a specific subset of topics was important to them personally, just 18 percent stated that the scale balancing economic and environmental goals was "extremely important," a number that once again ranks low in both an absolute sense and a relative sense (virtually tied with defense spending at the bottom of the list).

Table 6.5
Percent of Respondents Meeting Various Criteria for Issue Voting

Cumulative criteria for issue voting	Services/ spending	Abor- tion	Aid to blacks	Defense spending	Environment/ economy
Placed self on issue scale	85.5%	99.4%	91.1%	86.4%	85.2%
Placed self, Clinton, and Dole on issue scale	79.9	80.9	78.6	76.8	75.1
Placed self and saw difference between Clinton and Dole on issue scale	71.6	66.6	63.8	65.6	57.8
Placed self and saw Clinton as more liberal than Dole on issue scale	61.8	57.9	57.2	45.4	46.0
Placed self and saw Clinton as more liberal than Dole on an issue respondent thinks is "extremely" important	19.4	19.1	14.1	9.9	9.6

Source: American National Election Study (1996).
Note: For question wording on seven-point issue scales, refer to the appendix.

To demonstrate the combined power of these first two factors—low issue salience on the one hand, and an inability to perceive strong differences between the candidates on the other—table 6.5 summarizes a series of logical conditions necessary for issue voting.[30] At the most basic level, of course, respondents must be familiar enough with an issue to place themselves comfortably on a seven-point scale. While nearly all of those polled by the NES met the first requirement, steadily smaller numbers were able to place themselves and both major-party candidates for president on the same scale, fewer still in a way that identified any accurate difference between them. In the end, among those who could, fewer than 10 percent considered the issue of jobs and the environment to be "extremely important," once again ranking lowest among a list of five topics. Especially in acknowledging that those final figures represent not the act of issue voting itself but rather its realistic *potential*, the environment fares poorly indeed.

One final and complicating factor needs to be considered as well. Perhaps environmental issues fail to impact the vote for president because of the tendency of those concerns to cut across traditional (and more powerful) cleavages, including partisan identification. To address that possibility demands that candidate and party effects be isolated from the impact of the issue itself on vote choice. This is an undeniably difficult task. Still, in a similar study of abortion attitudes, Kevin Smith argues that issue voting can be disentangled successfully from other influences in cases where policy preferences and party loyalty are in conflict.[31] In other words, if an issue is politically potent, voters with the same party identification but opposite extremes of opinion should display different voting patterns.

To see if that expectation holds true, figure 6.1 visually plots the interaction of partisanship and environmental policy preferences in determining vote choice. Here, for the sake of simplicity, a postelection measure of the vote is viewed as a dichotomous preference between Bill Clinton and Bob Dole, with probabilities calculated from separate regression models run for each group of partisan identifiers. As a visual comparison of the slopes of each regression line makes clear, Republicans were more likely to vote for Clinton when their position on the environment/economy scale favored environmental issues. Despite the role of partisanship in shaping and filtering political information, Republican voters were not blind to Clinton's environmental stance, especially when that issue was one they felt strongly about.

Yet neither were voters especially inclined to cast ballots in the end that opposed either party simply out of environmental concern. When different extremes of opinion are compared, it is striking how willing Democrats were to vote for Clinton regardless of their environmental preferences. Among Republicans who strongly supported the environment, the issue had somewhat greater effect, lifting their probability of voting Democratic, but ultimately falling short. Why? Because the true impact of environmental preferences on the vote depends on intercepts— that is, on the starting points of individual voters. In this case Republicans were, *a priori*, more predisposed against Clinton for other reasons.

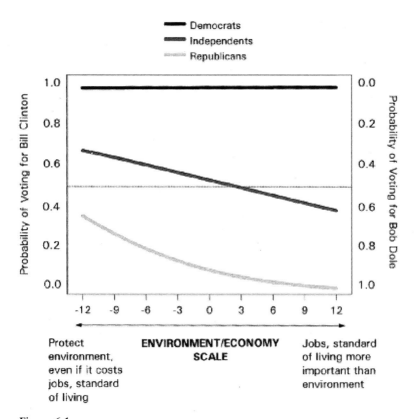

Figure 6.1
Partisanship and the Environment in the 1996 Presidential Election
Source: National Election Study (1996).
Note: By combining responses to variables v960523 and v960525, the independent variable used in this graph indicates both the direction and intensity of preferences along the environment/economy scale. See appendix for question wording.

His environmental positions may have made him a somewhat more attractive candidate to moderate Republicans, but in this case it appears not to have persuaded many to cross party lines.

The true impact of environmental issues, then, seems to lie not with partisan voters but rather with self-proclaimed Independents. For them, the environment may have been a crucial issue in 1996, one that pushed moderates in favor of casting a vote for Clinton when and if they prioritized environmental protection over the state of the economy.[32] Environmental concern may never supplant the deep anchor of partisanship, but for "swing" voters less attached to party, it may provide an important source of differentiation between candidates, if those policy differences are clearly articulated. In some elections and under some conditions, therefore, the electoral impact of environmentalism might be strong enough to cause a meaningful shift at the margins.

The 2000 Presidential Campaign

If environmental issues had yet to take center stage in a national campaign by 1996, the race for president four years later offered environmentalists a hint of something more promising. The Democratic candidate was Al Gore, President Clinton's heir-apparent, a self-styled environmentalist, and the author of *Earth in the Balance* (1992). But despite Gore's credentials and his past willingness to use the bully pulpit to champion environmental causes, political observers were struck largely by the quiet irony of Gore's campaign. Under fire on one side from environmentalists who believed that Clinton administration policies had not gone far enough and on the other by critics who labeled the views expressed in his book extremist, Gore allowed political caution to temper his environmental positions, relegating the issue to "the margins of his race for the White House."[33]

The Republican nominee, Governor George W. Bush of Texas, son of the man who in 1988 had pledged to be "the environmental president," made promises of his own in a series of speeches delivered at national parks in the Pacific Northwest. But the issue never quite emerged, as some hoped it would, as the "sleeper" issue of the presidential campaign.[34] What attention the environment did receive seemed focused instead on Ralph Nader's run for the White House as a third-party candidate, and his dogged pursuit of at least 5 percent of the popular vote—enough to secure federal funding for the Green Party in 2004. That Nader's support ultimately fell well short of his goal (just 2.7 percent of the vote nationwide) suggests that the environment, once again, failed to live up to its potential.

While the National Election Study did not replicate in full the environmental battery used in 1996, data for the year 2000 offer at least some perspective on the campaign and its candidates.[35] First, despite the presence of both Nader and Gore in the race—men whose political reputations were based, in no small part, on their environmental records—there is little evidence to suggest that the salience of environmental issues was any higher in 2000 than it had been four years earlier. When asked, just twenty-three respondents cited environmental issues as the single "most important problem" facing the nation. Amounting to less than 3 percent of the total, that statistic is virtually identical to the one obtained in 1996.

Measuring salience in more specific ways is difficult but no more favorable. Respondents in 1996 were asked to rate how important the issue of jobs and the environment was to them (only 18 percent considered it to be "extremely important"). In 2000, participants were asked the same question but in relation to environmental regulations placed on business instead. Without identical question wording, comparison between the two may be little more than speculative. In absolute terms, however, it can be said that the numbers remained weak, with just 12 percent believing that the subject was "extremely important."

Yet while the salience of the environment remained ironically low in 2000, keen divisions on policy did emerge between the two major-party candidates. Nader may have complained bitterly throughout the campaign that there was little substantive difference between Bush and Gore on the issues,[36] but in reality, as the editorial desk at the *New York Times* pointed out, "the Texas governor and the vice president offer[ed] as stark a choice on the environment as was ever put on view in a presidential contest."[37] As figure 6.2 visually demonstrates, the average difference that NES respondents perceived between the candidates on jobs and the environment was greater relative to the pairing of Clinton and

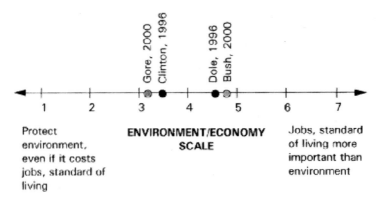

Figure 6.2
A Comparison of Candidate Issue Positions on the Environment and the Economy in the 1996 and 2000 Presidential Campaigns
Source: National Election Study (1996): v960526, v960529 and National Election Study (2000): v000714, v000719.
Note: See appendix for question wording.

Dole in 1996. Without the cognitive spark that salience provides, however, policy differences appear to be a necessary but hardly sufficient condition for voting green.

Finally, the weight of partisan tradition, as in the past, also played a role in shaping (and in some ways limiting) the outcome of the 2000 presidential campaign. While most Democrats and Republicans remained loyal to their preferred party regardless of their environmental preferences, Independents found Gore a less attractive candidate than Clinton overall for a multitude of reasons that had little to do with the environment. Voters in 2000 also continued to adhere to well-established voting patterns that consider ballots cast for a third-party candidate to be a "wasted" vote, leading many of those who were sympathetic to Nader to abandon him in the end because of what he later termed the "cold-feet factor."[38]

Given the benefit of hindsight, the greatest impact that the environment had on the 2000 presidential campaign may have been largely unintended. With a protracted outcome determined by an amalgam of state results in the Electoral College—with some help from the U.S. Supreme Court—Nader's small fraction of the vote failed to achieve his goal, but it undoubtedly contributed to Gore's hairbreadth loss. Playing the role of "spoiler," a shift of just one-half of 1 percent of Nader's support in Florida alone could have given Gore the state's twenty-five electoral college votes and with it the presidency itself.[39] From the standpoint of the mainstream environmental movement, then, it is not hard to imagine why Nader's candidacy was considered by some to be a "misguided crusade."[40]

Conclusions

Today, public concern for environmental quality stands as one of the most impressive findings in survey research. Public opinion polls show that the environmental movement has earned the sympathetic support of large majorities of Americans, many of whom claim the label *environmentalist* as their own.[41] Yet as Riley Dunlap notes: "In democracies, the 'bottom line' for judging the strength of public opinion is the impact of that opinion on the electoral process."[42] If that is the case, the news for environmentalists is both good and bad.

Environmentalists should take heart in the knowledge that environmental issues can, for certain voters under certain conditions, influence ballots cast in presidential elections. Although the numbers may be small, politicians who promote their environmental positions appear to do so wisely. As Christopher Bosso points out, "Strong public concern does not translate automatically into policy responses. It translates only into opportunities for leadership that may or may not be exploited," a lesson that Democrats are just beginning to learn.[43] In close races where victory is won at the margins, elite agenda setting by environmental candidates might well be successful in defining a unique environmental agenda and in increasing the salience of environmental issues from the top down—both factors that ultimately shape the likelihood of a "green" vote.

But as Al Gore's campaign for the presidency in the year 2000 demonstrates, there are dangers and limitations here as well. If the potential for environmental issues to make a difference at the polls lies mainly with "swing" voters who are less weighted by the anchor of partisanship, political parties and their candidates sympathetic to environmental issues may find themselves in a Catch-22. Independent voters are, in most cases, squarely in the middle of the road on matters of environmental policy and remain so in their general ideological views, so much so that candidates who promote policy differences with their opponents might risk being seen as too liberal or too extreme—in effect estranging the very voters they seek to attract.

Promoting policy differences, then, is risky, but it is also difficult. As Jedediah Purdy notes, "Although nearly everyone cares about clear air," the issues Vice President Gore embraced and at which he best succeeded tended to be "conducted in the language of parts per million, emission-control technologies, and cost-benefit analysis."[44] In other words, they were the prototypical "hard" issues that Carmines and Stimson say make issue voting near impossible. With "no ancient redwoods or endangered seal pups to provide television images of what was at stake," Gore's laudable record did little to ignite public enthusiasm for the environment or for his candidacy itself.[45] Environmentalists may have criticized George W. Bush's experience as governor of Texas and doubted his commitment to environmental protection, but his speeches on the subject were emotive and "symbolically perfect" nevertheless.[46]

If a majority of voters are genuinely concerned about the environment, it stands to reason that few political candidates come out opposed, at least in principle and in rhetoric, to environmental policies.[47] But as Peter Bragdon and Beth Donovan warn:

If more candidates on both sides of the aisle tout environmental credentials, it may become more difficult for these groups to draw public distinctions between allies and adversaries. As long as candidates like George Bush can win while touting environmental credentials that were highly suspect in the environmental community, politicians may have little incentive to change their behavior.[48]

In other words, given elections that invite environmental symbolism and the "greenwashing" of legislative records, pro-environmental candidates do not necessarily become pro-environmental presidents.[49]

With or without a strong contingent of active voters, however, there are other potential ways in which the environmental views of the electorate are represented. Environmental organizations such as the Sierra Club and the League of Conservation Voters serve as active mediators between citizens and their elected representatives. By lobbying Congress directly and by contributing funds to the political campaigns of pro-environmental candidates, these groups may help to hold representatives responsible for their legislative actions when and if voters fail to do so.[50] To suppose, then, that elected officials cannot be held accountable for their environmental policies without a tangible electoral incentive underestimates the ability of environmental concerns to influence the public agenda in other ways. As one columnist writes,

Without ever having elected a Green Party candidate to major public office or putting major components of their agenda on a ballot, environmentalists have succeeded—through agitation, litigation and cajoling friends in high places—in seizing the levers of power and bending the machinery of government to their will, turning the movement outside in.[51]

Finally, since environmental issues are seldom raised in presidential campaigns, criticism of the environment as a political issue may also reflect a continuing misperception of the nature of national elections. As Michael McCloskey bluntly puts it, "They are not plebiscites on this question."[52] Perhaps, then, we would more fairly judge the environment's "bottom line" by looking to other political arenas, such as ballot initiatives and referendums at the state and local level where environmental issues enjoy greater salience and less competition for room on a crowded political agenda.[53] If a stronger and more direct electoral connection can be found there, it may suggest that the uphill battle faced by environmentalists is at least half won.

7

The Ballot Box II: Environmental Voting on Statewide Ballot Propositions

Environmental ballot propositions are often considered bellwethers of the nation's willingness to take action on environmental issues, and of the extent to which U.S. consumers and taxpayers are ready, willing, and able to pay for costly environmental reform. If the environment's bottom line is truly to be judged by its impact on the electoral process, then, statewide ballot questions appear to present a golden opportunity.[1] Not only do initiatives and referendums provide the purest form of issue voting that occurs in American politics,[2] the process consistently offers citizens greater chance to address environmental issues in a succinct and direct way, shielded from otherwise dominant influences on electoral choice, such as partisanship and candidate appeal.[3]

Yet with an inconsistent record of election day victories and defeats over the years, scholars have interpreted the environmental scorecard in very different ways. While some highlight selective success as proof that Americans are willing to take a stand on environmental issues, especially when elected representatives fail to do so,[4] others insist that high-profile losses are devastating to the political credibility of the movement itself.[5] As one observer bluntly put it immediately following the 1996 election, in which environmentalists spent millions of dollars to influence the outcome: "The environmental movement set out to be the mouse that roared, but all Americans heard was a squeak."[6]

Explaining the conditions under which environmental issues succeed (or fail) at the ballot box is an important but undeniably difficult task. While Laura Lake finds that environmental ballot measures in California "met with the same or slightly better rates of approval than their non-environmental counterparts" over a ten-year period from 1970 to 1980, no study has yet attempted a comprehensive review of *all* environmental ballot questions over an extended time frame.[7] That deficiency is redressed in this chapter—through analysis of an ambitious list of more than 370 environmental measures offered in forty-five states between the years 1964 and 2000 and then through an examination of a broader and more detailed database of ballot propositions offered in four states over the same time period. When combined, both methods allow the overall success rate of environmental proposals to be compared (often favorably) to other issues of social, political, and economic importance, while providing statistical control for a variety of cross-pressuring factors, including ballot mechanics and economic conditions. Finally, since the economy has proven to be an important determinant of the vote, as have other factors less accessible to measurement on an aggregate scale (including issue framing and campaign finance), those topics are explored here using survey data from two prominent and contrasting case studies—the first is a successful 1986 toxics initiative in California intended to protect drinking water supplies, and the second is an unsuccessful 1992 Massachusetts recycling initiative. Consistent with David Magleby's claim that "the side that defines the proposition usually wins the election," both cases demonstrate that the content of a campaign message (and not simply the media visibility money affords) may be key to understanding public willingness to pay for protective environmental policies.[8]

Direct Democracy and the Environment

At its core, referendum and initiative refer to procedures that permit voters to cast ballots directly on issues of law, policy, and public expenditure. With ancestry extending back to the Greek city-state, through New England town meetings at the time of the American Revolution, to the U.S. Progressive movement at the turn of the last century, direct democracy has evolved and expanded in the United States over the past hundred years. While there is no procedure for a national referendum, all states but Delaware now offer opportunities for citizens to cast ballots in some form on constitutional and/or statutory issues, while just over half provide a form of popular referendum or initiative whereby citizens directly challenge existing law or propose new policy.

With the growth of single-issue constituencies since the 1970s, this latter form of direct legislation has become an "integral strategy" to lawmaking that is used by political activists to broaden the public agenda.[9] "Time and again," write Hugh Bone and Robert Benedict, "the sponsors [of ballot propositions] have been ahead of the legislature" in proposing changes in liquor laws, welfare benefits, environmental protection, and government reform. Indeed, the success of many of these measures at the polls, often by impressive margins, "can be interpreted to mean that the legislature was unresponsive to a wide-spread desire" for policy change, allowing controversial issues to bypass government stalemate by facing the electorate instead.[10]

While environmentalists are clearly not alone in their use of ballot measures to combat legislative inaction, the popularity of such strategies in many states demonstrates a fundamental compatibility between direct democracy and the populist, grassroots orientation of the environmental movement. As such, ballots cast on environmental referendums and initiatives provide a valuable reading on the relative importance of the environment to American voters and taxpayers.

Voting Behavior on Ballot Propositions

While most citizens are unlikely to vote green when casting ballots for political candidates (at least at the national level), voting on statewide referendums and initiatives presents increased opportunity, as well as an entirely new set of challenges. For example, voters approach the mammoth task of evaluating complex legislation in the absence of usually dominant cues or economizing devices, such as partisan identification or candidate evaluations.[11] The dilemma faced by many voters is made even more difficult by the intimidating number of measures offered on many state ballots and by the technical language used in wording proposals.[12] This informational vacuum can lead to risk-averse behavior and negative voting,[13] despite the social desirability of many environmental concerns. As a result, scholars caution that voting behavior on ballot propositions is more likely to be unstable over the course of an electoral campaign and more susceptible to advertising and other political appeals as voters strive to bring order to chaos.[14]

While voting behavior on ballot propositions may represent "strikingly idiosyncratic" decision making, scholars have attempted to identify consistent factors that influence proposition success or failure, including proposition type, ballot position and voter fatigue, election schedules, and economic conditions.[15]

Proposition Type

First, many students of state referendums have noted that measures placed on the ballot by the legislature enjoy a higher rate of success than those proposed by citizen petition.[16] In examining data from twelve states from 1898 to 1992, David Magleby estimates that 61 percent of all legislative propositions were approved by voters, while just 37 percent of those proposed directly by the people succeeded on election day.[17] This disparity seems to suggest that voters are more likely to trust the merits of legislative proposals over those offered by special interest groups or lobbyists, and to accord elected representatives the "benefit of the doubt."[18] A similar observation has been made with regard to constitutional versus statutory proposals. Eugene Lee argues, for instance, that voters are more likely to approve changes in policy rather than changes in state constitutions given that the latter is thought to be more fundamental and permanent.[19]

Ballot Position and Voter Fatigue

Second, in the absence of party and candidate cues that help to reduce information costs for voters, ballot position effects may be more likely as citizens make use of the limited information contained within the ballot itself to aid in vote choice.[20] Yet while some research shows that the relationship between ballot position and voter approval is "negative and statistically strong,"[21] and that proposals appearing at or near the top of the ballot "have a distinct edge,"[22] results in this area have been inconsistent, and at times contradictory.[23]

Election Schedules

Third, some evidence suggests that ballot propositions fare better in presidential-year elections. Voters may be more friendly to measures during these campaigns because of a heightened sense of issue competition and interest.[24] In addition, more proposals may be offered in those years if sponsors determine that their signature-gathering efforts will generate more "bang for the buck" by way of voter interest and turnout.

Economic Conditions

Finally, given the complexity of many ballot issues and voters' tendency to rely on easily accessible information when making vote choices, Shaun Bowler and Todd Donovan find that prevailing economic conditions are key.[25] They argue that the willingness of the electorate to adopt new policies, particularly bond measures that allocate public funds or incur indebtedness, is weaker when state economic conditions are poor.

Environmental Ballot Propositions, 1964 to 2000

Gathering data on statewide ballot propositions is an undeniably difficult task. As many scholars have noted, a comprehensive archive of referendums and initiatives does not exist, even though it is believed that well over three thousand have been voted upon in state elections, of which at least fifteen hundred were by citizen initiative.[26] Virginia Graham's detailed analysis of initiatives is a fine attempt at recording data on one type of ballot proposition, but is now itself nearly two decades out of date, and while newly supplemented by an extensive database at the Initiative and Referendum Institute,[27] no comparable study has yet been made of measures proposed by state legislatures. Moreover, Austin Ranney argues that "the disarray of some states' records makes such a compilation both costly and time consuming."[28] Given these difficulties, it is not surprising that most studies that focus on ballot propositions as a unit of analysis are limited by time, state or geographic region, or type of ballot measure considered, and frequently by all of the above in some combination.[29]

To obtain data for this chapter, complete lists of ballot questions offered in statewide elections were requested from election officials in all forty-nine applicable states over an extended length of time. In all, usable election results were received from forty-five states for the years 1964 through 2000, although some lists contained incomplete information on vote totals, while others failed to include recent election results. Whenever possible those gaps were supplemented by data drawn from state blue books or other published materials.[30]

The next step involved examining those lists in detail to code for measures with significant environmental content. That process was likewise difficult for several reasons. First, the quality of information provided by state election offices was inconsistent at best. While some lists contained lengthy descriptions of each ballot measure drawn from official election guides, others included little more than a short word or phrase that often required additional research. Second, subjective judgments had to be made regarding the definition of an environmental proposal. While most choices were clear, some were not, leading to the adoption of a rather strict standard. A simple example illustrates the point. Many referendums conducted during the mid-1970s proposed an increase in gasoline taxes. While those efforts could be viewed as an incentive toward the conservation of nonrenewable natural resources, most were not directly environmental in their intent but were rather designed to address an OPEC oil crisis abroad and for that reason are not included here. For similar reasons, measures funding public transportation as well as those regulating hunting, fishing and trapping are also excluded, despite in the latter case an obvious impact on efforts to protect certain species of wildlife. In the end, only those measures that could clearly be perceived as "environmental" by voters, without mixed messages and confounding factors, and without extensive knowledge drawn from lengthy voting booklets, were recorded.[31]

Trends in Environmental Ballot Measures

In all, based on the standards outlined above, more than 370 ballot measures on environmental subjects were offered to American voters between 1964 and 2000, and as figure 7.1 demonstrates, that number has grown over time (somewhat erratically), in large part due to the increased popularity of citizen-initiated forms of legislation. As figure 7.2 further shows, approval rates for environmental ballot measures appear to be as good as, if not better than, those for all other subjects. Recall that in sampling data from twelve states between 1898 through 1992, David Magleby finds that 61 percent of all legislative proposals succeeded at the ballot box, while just 37 percent of those proposed by citi-

Total number of environmental ballot propositions

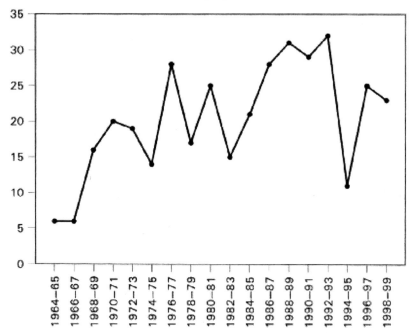

Figure 7.1
Environmental Ballot Propositions

zen petition were approved.[32] In contrast, 62 percent of legislative proposals on environmental subjects passed during our extended time frame,
as did 40 percent of all citizen-initiated referendums and initiatives. Despite recent high-profile losses that have garnered media attention, environmental ballot propositions generally fare quite well at the polls.

Aggregate statistics do reveal some degree of inconsistency and fluctuation in the number of such measures that are approved and rejected by
voters in any given year, however. For instance, while a majority of environmental ballot measures are approved in most elections, those topics
fared rather poorly (falling at or below a 50 percent success rate) in
1976, 1982, 1990, and 1992. There seem to be at least two possible
explanations for that pattern. Consistent with the argument made in
chapter 3, it may be that weak economic conditions influence election
outcomes and that voters hesitate to pay for costly environmental reform

Total number of environmental ballot propositions

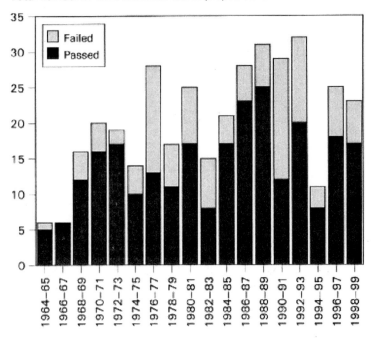

Figure 7.2
Environmental Ballot Propositions by Outcome

when faced with economic difficulties.[33] A second possibility also arises in noting that, with the exception of 1982, a higher than average number of citizen-initiated proposals were offered in those years (25, 12, 26, and 23, respectively). Given that initiatives typically face greater public opposition, election day disasters in those years may be little more than coincidental. To identify and disaggregate these (and other) competing influences, a more sophisticated model is required.

Voting Behavior on Ballot Propositions in Four States
One central issue to be explored here is simple: how much of the environmental movement's success at the ballot box can be attributed to fortuitous ballot position, favorable economic conditions, and the like, and how much can be attributed to the popularity of the subject matter itself? Conversely, how much of the environment's failure at the polls is

the result of outright bad timing rather than the actual merits of the proposals? To view referendum voting as "issue voting" demands that certain key factors (both internal and external to the ballot) be statistically controlled. It also requires some base line against which to compare the success rate of environmental measures relative to other issues. With that in mind, the most appropriate approach is to use multiple regression analysis to explain election outcomes on both environmental and non-environmental ballot propositions, distinguishing between the two with the use of a dummy variable.[34]

Given that an enormous volume of referendums and initiatives have appeared on state ballots over the past three decades and that environmental matters constitute (at most) 10 percent of those proposals, an analysis that attempted to compare all environmental measures alongside their non-environmental counterparts would be prohibitively large and time-consuming.[35] For that simple and pragmatic reason, complete election data from just four states between 1964 and 2000 are used here for more refined analysis.

To select states that would be broadly representative of the frequency and form of ballot propositions in the United States, case selection was geared toward maximizing variance along a wide range of considerations, including geographic location, population density, political ideology, and the overall use and popularity of referendum and initiative procedures. Given those criteria—as well as the always present need for high-quality available data—the following states were chosen for this sample:

• Colorado, which is largely Republican in its politics and known most recently for its controversial right-wing initiatives;

• Massachusetts, which while traditionally Democratic, shifted somewhat politically in the 1990s in the face of serious economic challenges;

• Oregon, which is a prominent center of environmental controversy in the Pacific Northwest due to, among other things, continued conflict between the protection of the spotted owl and the forestry industry; and

• South Dakota, a sparsely populated state that has experienced great economic fluctuation over the past thirty years due to its dependence on industries like agriculture and mining.

Using a data set of nearly five hundred ballot propositions offered to voters in these four states, regression results for a model of referendum voting are presented in table 7.1, where the aggregate percentage of votes cast in favor of each measure is regressed on a series of statistical controls, including the nature and source of the proposal, its position on the ballot, and the year in which it was offered. Measures of state and national economic conditions are also included. The results this model suggest several important conclusions.

First, the importance of certain ballot mechanics is confirmed. Measures proposed by citizen initiative or petition, for example, receive markedly less support. Second, while changes in personal income either at the state or national level have little impact on aggregate vote totals, the rate of unemployment has a strong, negative, and statistically significant effect in both areas. While each of these results replicates and supports previous work on referendum voting, the dummy variable differentiating environmental proposals from all other issues is most important for our present purposes. With a slope coefficient that is positive but statistically indistinguishable from zero, results confirm that environmental issues, on average, do at least *as well as* other types of social, economic, and political subjects, holding all else constant. Once again, media attention to high-profile losses seem at odds with actual election results, underestimating the strength of the environment as a ballot issue.

A Closer Look at Environmental Issues on State Ballots

If environmental measures, on average, tend to do as well as other types of issues on election day, do differences among environmental subjects help to explain why environmental voting on referendums and initiatives has traditionally been labeled inconsistent? A disaggregation of environmental proposals by topic in figure 7.3 suggests that they do.

In this case, a complete list of environmental ballot propositions is used, where the dependent variable represents the aggregate percentage of votes cast in favor of each question. After controlling once again for election year, ballot mechanics, and economic conditions, as shown in table 7.2, certain types of environmental issues do fare better at the polls than others. Bottle bills and other efforts at recycling, on average, receive aggregate votes that are nearly nine percentage points lower than the

Table 7.1
Proposition Voting in Four States, 1964 to 2000

Independent variables	OLS Slope coefficient	Standard error
Environmental proposal (0 = no, 1 = yes)	0.42	2.58
State (dummy variables):		
Colorado	3.63	2.49
Massachusetts	8.27**	3.12
Oregon	3.15	3.11
South Dakota	0.00	—
National economic conditions:		
Annual unemployment rate, seasonally adjusted	−2.63**	0.90
Rate of inflation as measured by the Consumer Price Index	−0.19	0.32
Percentage change in per capita disposable personal income	0.41	0.64
State economic conditions:		
Annual unemployment rate, seasonally adjusted	−1.07**	0.75
Percentage change in per capita disposable personal income	0.29	0.26
Proposition type:		
Proposed change (0 = statutory, 1 = constitutional)	2.53	1.94
Source of proposal (0 = state legislature, 1 = citizen petition)	−4.11*	2.07
Ballot position	−0.80	0.58
Ballot position^2	0.02	0.03
Presidential election year (0 = no, 1 = yes)	−1.67	1.58
Trend $(t_1, t_2, \ldots, t_n)$	0.08	0.10
Intercept	56.41	5.56

R-square = 0.13
Number of cases = 495

Notes: $^*p < .05.$
$^{**}p < .01.$
$^{***}p < .001.$

Total number of environmental ballot propositions

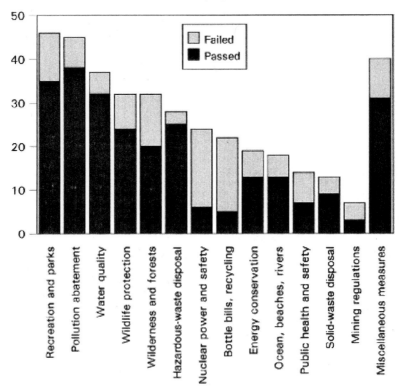

Figure 7.3
Environmental Ballot Propositions by Subject

base category, while efforts to protect public health and safety through pesticide regulation, asbestos removal, or product labeling earn votes that are nearly thirteen percentage points lower. Measures designed to regulate nuclear power and safety also face significantly greater public opposition than other environmental causes. Why might these measures, in particular, fail?

Nuclear Regulation   First, despite a high degree of suspicion and concern, voters have reacted in inconsistent ways to measures regulating nuclear power and safety over the past thirty years, with just 25 percent of propositions on the subject approved by voters between 1964 and

Table 7.2
Explaining Variance in the Aggregate Vote on Environmental Ballot Propositions, 1964 to 2000

Independent variables	Slope coefficient	Standard error
Environmental subject matter:		
Pollution abatement	−2.63	2.79
Wildlife protection	−0.43	3.04
Wilderness, forests, and open space	−7.77*	3.00
Ocean, beaches, and rivers	−3.96	3.53
Parks and recreation	−6.86*	2.81
Energy conservation	−7.43*	3.66
Nuclear power and safety	−7.58*	3.60
Public health and safety (e.g., regulating human exposure to toxic substances)	−12.47**	3.96
Water quality and conservation	−0.50	2.95
Solid-waste disposal	−6.22	3.91
Mining regulations	−4.32	5.20
Hazardous-waste disposal	4.53	3.19
Bottle bills, packaging, and recycling efforts	−8.98*	3.67
Miscellaneous subjects	0.00	—
National economic conditions:		
Annual unemployment rate, seasonally adjusted	−0.49	0.62
Rate of inflation as measured by the Consumer Price Index	−0.29	0.30
Percent change in per capita disposable personal income	0.39	0.51
Proposition type:		
Proposed change (0 = statutory, 1 = constitutional)	2.46	2.13
Source of proposal (0 = legislature, 1 = citizen petition)	−11.44***	2.14
Bond issue (0 = no, 1 = yes)	1.12	2.03
Presidential election year (0 = no, 1 = yes)	−1.21	1.39
Trend $(t_1, t_2, \ldots, t_n)$	−0.07	0.10
Intercept	66.85	5.77
R-square = 0.34		
Number of cases = 349		

Notes: $* p < .05$.
$** p < .01$.
$*** p < .001$.

2000. While Betty Zisk argues in *Money, Media and the Grassroots* (1987) that the nuclear accident at Three Mile Island in 1979 probably influenced voters' perceptions of nuclear safety, leading to the electoral success of ballot questions in some states that had previously been defeated by substantial margins, economic arguments regarding loss of jobs and rising electrical rates for consumers remain potent political weapons used by opponents to defeat regulatory efforts.[36]

Public Health and Safety   Second, efforts to protect public health and safety through pesticide regulation, asbestos removal, product labeling, and the like often face stiff opposition at the polls despite appealing ballot titles and seemingly intense national concern. While the use of some pesticides is known to have carcinogenic and reproductive effects at high concentrations, for example, their use is often balanced in political debates by tangible agricultural benefits that maintain high crop yields, attractive cosmetic standards, and stable market prices. Moreover, despite the inherent appeal of "right-to-know" campaigns that promise to alert the public of exposure to toxic chemicals, potentially high costs of compliance for business and industry lead to complaints that these measures harm competitiveness and employment.

Bottle Bills and Recycling Efforts   Finally, "bottle bills" mandating deposits on bottles and cans and other recycling efforts designed to regulate product packaging have traditionally faced opposition from the beverage industry, as well as voters in many states. Even though states with deposit-refund systems boast higher than average recycling rates and reduced curbside litter,[37] opponents argue that such measures lead to job losses and increased prices for consumers, while offering marginal environmental benefits, at best.[38]

With that additional background in mind, data results presented here suggest that economic conditions remain a persuasive factor in referendum voting. While the national rate of unemployment has a strong, negative impact on voting on all issues, the environmental measures that fail most frequently seem particularly sensitive to charges of economic cost and job loss. To better understand the complex relationship between the economy and the environment in the eyes of the American voter, it is necessary to refine our focus one last time by concentrating on two case studies that fall squarely into the typology developed above—one that succeeded at the polls and one that failed.

Case Study 1: The California Toxics Initiative, 1986

In a state well known for its liberal support of environmental policies, continued concern over the safety of public drinking water supplies and ongoing frustration with legislative inaction encouraged environmental groups to offer a unique solution to California voters during the 1986 general election. Proposition 65—formally known as the Safe Drinking Water and Toxic Enforcement Act of 1986—was designed to restrict the release of toxic substances in "significant amounts" into drinking water if those chemicals were known to cause cancer or birth defects. Moreover, the initiative required California companies to give "clear and reasonable" warning before knowingly exposing the public to harmful chemicals from a variety of sources, including alcoholic beverages, paint, dry cleaning fluids, and gasoline. While product labeling was most clearly intended to inform the public of risky products and activities, environmentalists also believed that "the greatest impact" of the initiative could be on manufacturers rather than consumers, by "prodding them to reformulate products and get hazardous chemicals out of the environment and the workplace," which would avoid the need for warning labels altogether.[39]

Finally, and perhaps most significantly, in permitting citizen lawsuits to be filed against offending companies, Proposition 65 promised a dramatic "change in the rules about who is responsible for setting safe chemical exposure levels."[40] Rather than relying on a massive state and federal bureaucracy to study scientific evidence and determine safe levels of exposure—a process that is often excruciatingly slow—Proposition 65 offered an "innovative legal approach" that effectively turned the regulatory tables by placing the burden of proof on businesses to show scientifically that the chemicals they use are safe.[41]

This latter requirement, in particular, solidified a fierce opposition group to the initiative that included the oil industry, utility companies, the California state Chamber of Commerce, the Farm Bureau Federation, and the California Manufacturers Association, as well as incumbent California governor George Deukmejian. In outspending environmental proponents by a three-to-one margin, "No on 65" groups accumulated a war chest of nearly $5 million and went so far as to run a full-page ad in the *Wall Street Journal* urging companies nationwide to help them "prevent the *second* biggest business disaster in California history," in their opinion, ranking only behind the stock market crash of 1929.[42]

In presenting their case before California voters, opposition groups argued that prohibiting toxic discharges in "significant amounts" (defining *significant* as "any detectable amount") would essentially ban the use of many chemicals useful to agriculture and industry, given the sophistication and sensitivity of current scientific technology.[43] The "No on 65" campaign also argued that the initiative was blatantly unfair and "full of exemptions," granting immunity to government agencies while creating still more bureaucratic red tape for private sector businesses.[44] As Deukmejian noted: "If [the initiative] is genuinely concerned, as proponents say it is, with the public health, then it shouldn't matter" whether that risk comes from business or government. "This isn't an anti-pollution initiative," he argued; "it is an anti-business measure" that would place an "unbearable burden" on California businesses and farmers, while exempting large well-known polluters like the City of Los Angeles.[45] In sum, opponents argued that Proposition 65 would harm the California economy, drive away jobs, and not "result in one single glass of cleaner water."[46]

Understanding Public Attitudes toward Proposition 65
Between July 24 and October 30, 1986, the Field Institute (a nonpartisan public-policy research organization that collects public opinion data on a variety of social and political topics within California) conducted a series of three polls measuring public attitudes toward the upcoming state election, its candidates, and its ballot propositions.[47] Respondents in all three surveys were asked whether they had "seen or heard anything about an initiative, Proposition 65, that will be on the November statewide election ballot having to do with toxic substance[s] and drinking water." Voters who were aware of the measure were then asked how they might vote based on the information they had already received. All

Percent responding

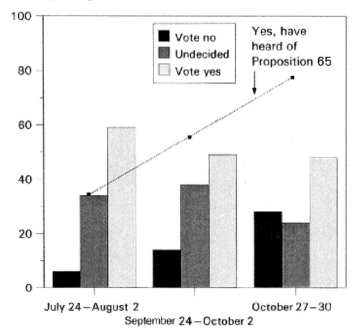

Figure 7.4
Vote Intentions among Informed Voters for Proposition 65
Source: Field Institute, California Polls (86-04, 86-05, 86-06).

participants were then read a brief neutral description of the initiative
and asked one final time how they would vote if the election were being
held that day.

Taken together, data results from these three surveys demonstrate
strong public support for Proposition 65. During the summer prior to
the election, 88 percent of those responding supported the initiative after
hearing a brief description of its goals. By the beginning of October, well
into the fall campaign and just one month prior to election day, overall
support remained high at 79 percent. But as figure 7.4 demonstrates, as
the campaign moved forward and information about the initiative became widely available, opposition among voters familiar with the measure increased substantially, from just 7 percent during the summer
months to nearly 30 percent by the end of October. The more voters
knew about Proposition 65, they more they disliked it.

Table 7.3
Factors Affecting Willingness to Vote for Proposition 65

Independent variables	Ordered probit coefficient	Standard error
Age	−0.01	0.00
Education	−0.09	0.07
Income	−0.02	0.04
Partisan identification	−0.00	0.04
Political ideology	0.10*	0.05
Race	0.28	0.27
Gender	−0.20	0.14
Prior knowledge of proposition 65	−0.62***	0.15
Intercept 1	1.61	0.36
Intercept 2	0.53	0.07
Log-likelihood = −284.615		
Number of cases = 447		

Source: Field Institute (September 24–October 2, 1986, n = 1023).
Notes: See appendix for question wording.
* $p < .05$.
** $p < .01$.
*** $p < .001$.

In the end, as table 7.3 confirms, strong support for the measure was sustained, in no small part, by the positive initial reaction of uninformed voters. Even after controlling for a variety of social, political, and demographic factors, including partisan identification and political ideology, voters who were unfamiliar with Proposition 65 (and who therefore had no knowledge of negative advertising aimed against it) were much more likely to favor the measure, probably because of its appealing goals to "prohibit the discharge of toxic substances into drinking water" and to "require warnings of toxic chemicals exposure." An equally appealing ballot title on election day also meant that voters heading to the polls with little or no knowledge of the initiative were likely to support it. As opponents feared, this strong "gut reaction" contributed to an impressive victory for environmentalists, where the toxics initiative attracted 63 percent of the popular vote. As one industry representative put it, given a margin of defeat of nearly two to one: "[W]e sort of got our clocks cleaned."[48]

Why Did Proposition 65 Succeed?

As a strict environmental policy placed before California voters, the importance of Proposition 65's success cannot be overestimated. It was, after all, the first fiercely contested environmental ballot measure to succeed in California since the 1972 Coastal Act,[49] and its emphasis on pollution prevention, rather than cleanup, was both innovative and resourceful given the lethargy and limitations of federal tolerance-setting. As a bona fide electoral success, however, Proposition 65 would undoubtedly be considered a deviant case by Zisk and others who stress the importance of campaign spending in determining initiative outcomes, particularly given high and disproportionate spending by opponents to the issue.[50] On these grounds, the success of Proposition 65 clearly demands attention, as well as explanation.

First, despite a well-funded campaign against the proposal, the message that opponents sent to California voters was clearly ill advised. The "exemptions" issue may have convinced some that the toxics initiative was either inadequate or unfair, but it was the source of that message ultimately failed to ring true. As one newspaper columnist put it, "the oil and chemical companies went on TV to attack the measure as too weak. Their slogan was: 'Too many exemptions,' as if *they* were the environmentalists."[51]

Second, as California poll data suggest, an appealing ballot title probably meant that many undecided or uninformed voters would cast ballots in favor of Proposition 65. In this case, broad environmental goals were described in an appealing manner that very likely encouraged support based on social desirability alone. This "standing opinion" was strengthened further by ballot wording that placed clear responsibility (and potential cost) on big business rather than on consumers themselves.

Finally, while well-financed opponents tried to label the toxics initiative as a "simplistic response to a complex issue," California voters ultimately preferred to view the proposal as a clear-cut health and quality-of-life issue rather than an intricate political debate over policy means.[52] While such rhetoric may ultimately weaken the sophistication and potential of issue voting on environmental concerns, this strategy is fundamentally compatible with the informational vacuum faced by many citizens in the voting booth.

Case Study 2: The Massachusetts Recycling Initiative, 1992

Just four years following a major environmental victory in California, election day in 1990 was a particularly black Tuesday for environmentalists. With voters heading to the polls just six months after the twenty-fifth anniversary of Earth Day, environmental ballot propositions—the largest number ever presented to voters in a single election—seemed likely to ride the crest of record high levels of environmental concern and awareness. Indeed, as one commentator recalls, "So strong was the apparent green tide that even many traditional opponents accepted it as inevitable."[53] By November, however, the green tide had been crushed to a mere ripple. Of the twenty-six environmental measures appearing on state ballots that year, just thirteen passed, prompting newspapers from coast to coast to carry stories highlighting elections results with colorful language such as "flopped," "massacre," and "mowed down."[54] It was, as one major newspaper put it, a string of "stunning electoral defeats."[55]

Equally stunning was the nature of many of the defeated measures, including efforts to ban cancer-causing pesticides, reduce emissions of greenhouse gases, preserve forest land and habitat for endangered species of wildlife, and levy taxes on the transportation of hazardous materials. A toxics initiative in Ohio patterned after California's popular Proposition 65 was defeated by a margin of three to one, as was an Oregon recycling initiative designed to regulate product packaging.

Two years later, with a similar recycling measure pending on the Massachusetts ballot, the political fortunes of the environmental movement remained uncertain. While traditionally liberal and Democrat in its politics, Massachusetts was recovering from an economic slump that year that had produced falling revenues and serious job losses, as well as a new Republican governor, William Weld. In placing Question 3 on the ballot, sponsors hoped that a new policy regulating the recycled content of all product packaging produced in the state would "close the loop" in the state's burgeoning recycling program by creating local markets for the old newspapers, milk bottles, and plastics collected curbside by town recycling programs.[56] Environmental leaders were equally optimistic that the recycling initiative would help to boost the sagging Massachusetts economy by creating new jobs and encouraging resource efficiency.

In endorsing the measure personally, even Governor Weld believed that the measure would "be good for the state's economy as well as its environment."[57]

Opposition to the initiative, however, came from several well-financed groups, including packaging companies, the plastics industry, the American Paper Institute, and the Grocery Manufacturers Association, which together formed a vocal "No on 3" coalition. Over the course of a $6.5 million advertising campaign that outspent environmental sponsors by a ratio of thirteen to one, opponents argued that consumers were likely to bear the burden of mandated repackaging, costing families up to $230 dollars per year in higher product prices.[58] They also contended that enforcement of the initiative would require a costly new state bureaucracy, burying grocery stores in bureaucratic red tape, while having little impact on the state's solid-waste problems, given that over 80 percent of Massachusetts packaging originates outside the state.[59]

The Voters Respond
How did Massachusetts voters react to Question 3? As figure 7.5 demonstrates, initial reaction to the proposal was overwhelmingly favorable. As Magleby would predict, however, support weakened considerably over the course of the election campaign, particularly after negative advertising began to appear in the mass media in mid-September.[60] In a series of polls conducted by the Boston firm of Marttila & Kiley, Inc., 88 percent of respondents said they were likely to support Question 3 in February of that year, while by October just 55 percent said they would do so. During that time, well-financed opponents successfully recast the recycling issue as a packaging ban, stirring voter concern for jobs and the economy and voter disapproval of government waste, inefficiency, and bureaucracy. By election day, Question 3 failed at the polls, winning just 41 percent of the popular vote.

Understanding Public Attitudes toward Question 3
Why did such a drastic reversal of opinion occur over an eight-month period? Perhaps the best explanation can be found in the February 1992 poll conducted by Marttila & Kiley.[61] In interviewing 402 likely Massachusetts voters, pollsters asked respondents how they might vote for the

Percent responding

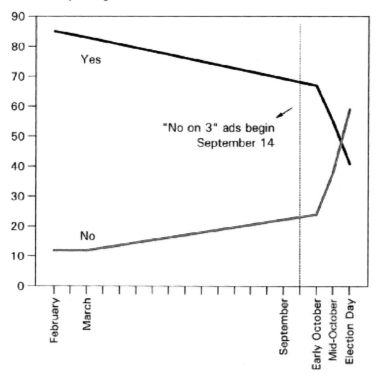

Figure 7.5
Trends in Support for the Massachusetts Recycling Initiative
Source: All data are provided by Marttila & Kiley, Inc. Election day data are
actual vote totals.
Note: Percentages may not add to 100. Remainder answered "not sure" or
"refused."

recycling initiative under three different conditions. First, respondents
were asked for their "initial reaction" to a proposal that "would require
nearly all packaging used in the state to be recyclable or made of recycled
materials." At a later point in the questionnaire, respondents were read a
more detailed description of the initiative and asked which way they
would be inclined to vote "if the election on this proposal were being
held tomorrow." Immediately following, respondents were asked to react to two batteries of questions—the first grading the importance of

various reasons to support the proposal and the second judging the merits of various reasons to reject it. A final measure of vote intention was used following discussion of these positive and negative attributes.

As data presented in table 7.4 make clear, a dramatic shift in opinion can be seen across these three measures due to framing effects that were designed to test the strength and stability of attitudes toward Question 3. While 88 percent of those responding were initially favorable to the proposal, with nearly all continuing to support the measure after hearing greater detail, just 53 percent of voters felt that they would "probably vote yes" after being exposed to the political arguments made by both environmentalists and their critics during the course of the campaign. In comparing responses to the second and third items, in particular, just 3 percent of voters were more likely to support Question 3 by the end of the survey, while 38 percent—nearly 2 in 5—felt that their support had weakened.

What factors most contributed to this downward shift? To compare vote intentions immediately before and after respondents were experimentally exposed to various political frames, a new variable was created by calculating the difference between the two measures. This scale ranges in value from -3 to $+3$, where 0 indicates no change in vote intention between the second and third survey conditions. The presence of a negative sign means that a respondent's support for Question 3 weakened, while a positive sign indicates that their vote intention strengthened. Using this measure of change as a dependent variable in table 7.5, several important conclusions can be drawn.

First, while political ideology has a strong impact on initial vote choice, it appears to have little independent effect on changing voters' minds either way during the course of the questionnaire, even though several criticisms made against the initiative in the survey were closely related to traditional ideological beliefs (such as taxation, bureaucracy, or regulation). Data also indicate that respondents who voluntarily recycled or who participated in town recycling programs were not more likely to support Question 3 in the end, despite the expectation that prior recycling behavior would have helped to invest Massachusetts voters in the outcome of Question 3 and in its efforts to create local markets for recycled materials.

Table 7.4

Three Variations in Vote Intention for Question 3

Voter Intention	Yes	Lean Yes	Lean No	No
Please tell me whether your initial reaction would be to vote in favor of that proposal or vote against the proposal: a proposal that would require nearly all packaging used in the state to be recyclable or made of recycled materials.	87.9%	—	—	12.1%
Let me describe this proposal in a little more detail. Packaging accounts for roughly one-third of the total volume of trash disposal in Massachusetts each year. In order to sharply reduce the amount of trash, this proposal would required that by July 1, 1996, packaging will have to meet one of five standards, by being smaller in size, reusable, recycled, or made of recycled or recyclable materials. Manufacturers and businesses can use any one of the five standards to meet the new packaging requirements. If the election on this proposal were being held tomorrow, would you be inclined to vote yes or no on this proposal? [If not sure:] I know you could change your mind, but which way are you leaning based on this information?	87.2	5.6	1.0	6.1
[After asking respondents to react to a lengthy list of positive and negative reasons for supporting or rejecting the proposal:] Now that you have heard some of the practical concerns and reservations about the recycling initiative, I want to see how you feel now: if the election were held tomorrow, would you probably vote yes on this proposal, are you leaning toward voting yes, are you leaning toward voting no, or would you probably vote no on this proposal?	52.8	29.4	7.2	10.6

Source: Marttila & Kiley, Inc. (February 11–23, 1992), $n = 402$.

Table 7.5
Factors Affecting Willingness to Vote for Question 3

Independent variables	Ordered probit slope coefficient	Standard error
Age	0.01	0.03
Education	0.02	0.06
Income	0.06	0.06
Partisan identification	0.08	0.12
Political ideology	0.13*	0.06
Gender	0.29	0.16
Household recycles	0.19	0.18
Additive scale of positive evaluations	0.03	0.02
Additive scale of negative evaluations	−0.06***	0.01
Intercept 1	−0.60	0.80
Intercept 2	0.05	0.02
Intercept 3	0.07	0.03
Intercept 4	1.08	0.11
Intercept 5	1.67	0.16

Log-likelihood = −243.193
Number of cases = 254

Source: Marttila & Kiley, Inc. (February 11–23, 1992), $n = 402$.
Notes: See appendix for question wording and scale construction.
* $p < .05$.
** $p < .01$.
*** $p < .001$.

Finally, and most important of all, it is clear that positive and negative frames had an asymmetric effect. While positive evaluations played little or no role in strengthening vote intentions toward Question 3, reservations expressed against the proposal were a major factor in eroding public support. A shift in just two categories on each item in the negative battery (for example from "minor reservations" to "strong reservations") leads to a statistically and substantively significant shift in vote intention, holding all else constant. When asked in an open-ended format which of the particular concerns mentioned they considered to be "most serious," respondents whose support declined commonly mentioned "loss of jobs" (29 percent) and "cost to consumers" (20 percent). If Massachusetts voters responded disproportionately to negative charges against the recycling initiative, therefore, they were particularly sensitive to these specific economic concerns—ones that were actively exploited by opponents throughout the initiative campaign.

Why Did Question 3 Fail?

Question 3 failed, in part, because although support for the measure in early polls was high, it was also soft. As Magleby argues, most voters begin a campaign knowing little, if anything, about the propositions they are likely to face on the electoral ballot, and so it is not uncommon to find widespread opinion change over time as voters gather more information about the issues.[62] Most commonly for initiatives, he says, that pattern is one of strong early support followed by defeat, especially when there is high spending on the negative side of a proposition. The failure of the Massachusetts recycling initiative clearly seems to fulfill these expectations.

But as Magleby notes, too, controlling the terms of debate can be key to understanding eventual success or failure. By reframing the issue at stake as a packaging ban rather than a pro-recycling policy, opponents were able to successfully shift public attention away from environmental concern toward potential economic costs that would presumably be born by Massachusetts consumers and their families.[63] The political context in which this shift occurred is also important in that it allowed opponents to simplify the issue at hand and to link public attitudes toward Question 3 to other factors, such as political ideology and long-standing beliefs about bureaucratic waste and inefficiency. Ultimately, its failure was not solely the result of lopsided campaign spending and negative advertising, although certainly those resources helped to inform Massachusetts voters of alternative positions. Instead, the "No on 3" campaign succeeded because its content and tone resonated with anxious voters.

Conclusions

Despite a series of notorious disasters in recent decades—including California's "Big Green"—where complex and innovative environmental proposals went down to electoral defeat, environmentalists continue to insist that setbacks at the ballot box do not indicate that public commitment to environmental protection has weakened.[64] Yet as Jessica Mathews freely admits, disappointing election results in the face of evident concern amounts to a troubling "discrepancy" that must be explained nevertheless.[65] The analysis presented in this chapter offers several possibilities.

First, despite high-profile losses that attract media attention and the vocal complaints of some that the environmental movement amounts to little more than a political "paper tiger,"[66] environmental referendums and initiatives do reasonably well at the ballot box. In comparing success rates for ballot questions on the environment to questions on all other issues, environmental proposals on average fare well once the nature and source of those measures are controlled. In other words, the failure of some environmental issues may be due to the controversial nature of the initiative process in general, rather than to public protests against environmental regulation itself. In this sense, it is important to recognize that the intensity of voters' feelings does not always reflect or necessarily represent their political preferences. "Direct legislation can," in David Magleby's words, "be a most inaccurate barometer of opinions," especially on issues where voters face a profound vacuum of information.[67]

In the absence of typical cues and informational shortcuts, Magleby and others argue that voters are more dependent on political campaigns to simplify choice and shape referendum decisions. Indeed, the power and potential of campaigns are clearly seen in the two case studies developed here, where in both instances political opponents used powerful rhetoric to try to persuade environmentally concerned voters that the marginal benefits of regulation failed to exceed its economic cost.

By linking the debate over product packaging to preexisting fears about state unemployment and bureaucratic red tape, political opponents to Massachusetts' Question 3 were able to redirect voters' attention away from the environmental benefits of the proposal and toward an acknowledgment of its costs in a way that ultimately undermined support for the proposal. While an ability to outspend environmentalists allowed business and industry groups to dominate the airwaves during the fall of 1992, the effectiveness of their attacks can be clearly seen in a poll conducted months before. In this sense, money becomes a proxy for other things that deserve equal attention.

Although money is instrumental (of course) in funding polls, campaign consultants and qualified staff, it cannot always compensate for a poorly constructed message. By choosing to emphasize the complexity of Proposition 65 in a multimillion dollar advertising campaign, opponents asked Californians to view a potentially "easy" issue regarding public health and safety as a "hard" policy debate, something overwhelmed voters were loathe to do.[68] It was a tactical mistake that allowed environmentalists to win an unlikely victory as financial underdogs. With a pro-environmental campaign in Massachusetts six years later that failed after making many of the same strategic errors, election results in these two cases seem to underscore the importance of simple and lucid proposals that sympathize with the substantial informational demands placed on voters.

In the end, evidence suggests that voting green on ballot propositions is more likely in simple, inexpensive, and low-key campaigns that avoid uniting political enemies.[69] As survey data from the California and Massachusetts elections attest, initial standing opinions on environmental issues remain overwhelmingly positive. Given that salience increases the likelihood of political opposition as well as the probability that voters will have been exposed to potent negative advertising, environmentalists unable to win the money war might fare better in the long run by promoting quiet, incremental reform rather than broad, sweeping changes in environmental law.

8

The Marketplace: Motivating the Citizen-Consumer

With an ambitious electoral record marred by legislative gridlock and the defeat of key environmental reforms in recent years, public concern has increasing found new outlets of expression outside the sphere of politics. Ever since the media success of Earth Day in 1990, American businesses have welcomed the "rise of a new consumer" who is motivated by increasing environmental awareness and a willingness to vote at the cash register, if not always the ballot box, for improved environmental quality.[1] In fact, this burgeoning demand for environmentally safe products, which exceeded $1.8 billion in sales annually by 1990,[2] was called "the political, economic, and social trend" of the nineties—a force powerful enough to build bridges and create unlikely alliances between environmentalists and industry.[3]

The apparent success of environmental efforts in the marketplace is surprising, however, when compared to the relative impotence of environmental issues in U.S. elections.[4] Political theorists have long recognized that decisions reached in the voting booth may differ from those made in the marketplace,[5] but private consumptive choice is typically thought to reflect narrow self-interest. In the political realm, as Cass Sunstein argues, "social and cultural norms often press people, as citizens, in the direction of a concern for others or for the public interest."[6] If those conventions hold, the fact that Americans seem more willing to buy green than to vote green, suggests an intriguing paradox that warrants careful attention.

Using survey data from a variety of sources, this chapter explores patterns of behavior among citizens and consumers relative to the environment. In examining why green consumerism outranks political actions such as voting, this chapter considers four different explanations—simple self-interest, political ideology, personal efficacy, and product advertising. In the end, in a world where politics and economics converge, those variables are of considerable importance, given that strong public support for environmental policies may be incapable of protecting the environment alone without an equal commitment in the marketplace.

A Paradox Defined

Economists and public choice theorists have long viewed voting behavior as a rational calculus made by thrifty, self-interested taxpayers concerned only with maximizing their own utility income.[7] Yet as Donald Philip Green notes, the empirical record in support of such a proposition is slim.[8] Time and again, he writes, statistical analyses find that "personal costs and benefits are rather poor predictors of how people want government to behave," either in the area of bilingual education,[9] affirmative action,[10] income redistribution,[11] social security,[12] or a host of other social issues and concerns.

A growing number of political theorists have suggested instead that public spiritedness and altruism serve as important components of mass political behavior.[13] In explaining why the calculus of decision-making might be different in the marketplace than in the arena of politics, James Buchanan notes that "a sense of participation may exert important effects on the behavior of the individual."[14] In contributing to a collective political outcome, he says, there is a rearrangement of the preference scale, one that encourages citizens to take a greater account of the public interest. In other words, with the promise of communitarian goals at the ballot box, we "adapt not only our actions, but even our desires."[15]

Yet to suppose that political and economic decisions really are different in character creates difficulty when we look to apply those behavioral expectations to issues like the environment. We might expect to see a decided shift in behavior as we move from the realm of private choice to that of social choice, and yet often we do not. We might believe, too, that private consumptive choices made by consumers in the marketplace would tend to reflect narrow self-interest, while electoral choices made by citizens in the voting booth would lean toward a broader concern for the public good. Instead, election results and market reports at times seem to plot that pattern in reverse.

While few citizens seem to cast ballots for presidential candidates on the basis of their environmental preferences, generating public support for statewide environmental initiatives and referendums has likewise been an uphill challenge for environmental groups, especially those outspent by well-funded opposition campaigns.[16] Sunstein insists that Americans "support government regulation that diverges from their behavior as consumers," but election results on ballot propositions in states like Ohio and Massachusetts suggest an entirely different story. Polls show that most Americans recycle regularly and are willing to pay more for products packaged in recycled materials, yet as chapter 7 points out, voters in those states overwhelmingly rejected initiatives that would have regulated the recycled content of product packaging.

Environmental concern seems to fare better in the marketplace. Today's eco-market represents a diverse collection of products and services, ranging from phosphate-free detergents to recycled paper products and rechargeable batteries.[17] While most studies find that deep commitment to the environment is concentrated in the hands of a privileged few—ranging from 5 to 25 percent of the U.S. population (depending on the stringency of the criteria used)[18]—public willingness to purchase certain environmental products, even at slightly higher cost, appears to run surprisingly deep. One survey in 1992 found that nearly three-quarters of consumers were "at least sometimes" influenced by environmental claims in the marketplace, and most appeared willing to pay at least 5 percent more for products known to be environmentally safe.[19]

Is the paradox of marketplace success and ballot-box failure simply an artifact of trying to compare apples and oranges—that is, of trying to compare in an anecdotal way behaviors that are essentially incomparable? Probably not. It is possible to contrast different forms of environmental behavior directly in public opinion polls, and although such efforts are rare, the results they produce are telling. In one survey, conducted by Cambridge Reports Research International in July 1993, respondents were asked to rate how frequently they participated in certain activities based on their concern for the environment. That list included multiple measures of consumer choice and one very impor-

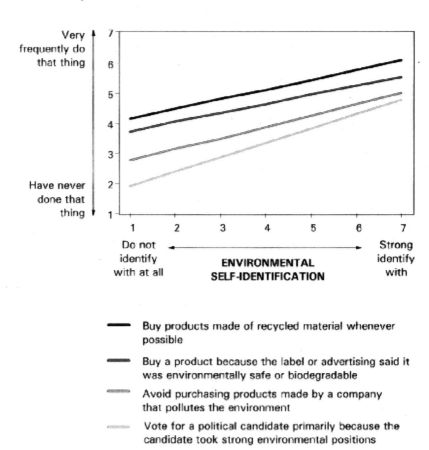

Buy products made of recycled material whenever possible

Buy a product because the label or advertising said it was environmentally safe or biodegradable

Avoid purchasing products made by a company that pollutes the environment

Vote for a political candidate primarily because the candidate took strong environmental positions

Figure 8.1
A Comparison of "Green" Voting and "Green" Consumer Behavior
Source: Cambridge Reports Research International (July 15–27, 1993).
Note: See appendix for question wording.

tant measure of environmental voting behavior. Consistent with what scholars have suspected regarding the likelihood of the latter, voting for a political candidate "primarily because the candidate took strong environmental positions" ranked low on the list overall. In fact, as a series of regression lines plotted in figure 8.1 demonstrate, voting ranked lower across the board, even among those who strongly identified with the term *environmentalist*. Americans, it seems, simply prefer to buy green rather than vote green.

In Search of Answers

Why do environmental issues seem to violate a reasonable set of behavioral assumptions regarding choices made in the marketplace and in the voting booth? Why should green consumer behavior be considered a powerful social and economic force, while identical environmental issues languish in electoral campaigns? A fully satisfactory answer is not immediately apparent. An individual's intensity of concern for the environment cannot be a determining factor since it holds constant across different kinds of behavior at any given point in time. Other easy answers fail to ring true as well. Aside from information costs, voting for political candidates with similar environmental viewpoints would appear to impose little marginal cost on a voter already heading to the polls on election day. Purchasing environmentally safe products in the marketplace, meanwhile, often demands a premium price tag, as well as a deviation from brand loyalty.[20]

Perhaps, then, it is because environmental issues must compete for attention alongside other priorities during crowded election campaigns, and yet the environmental attributes of a consumer product, such as paper towels or dish detergent, are similarly experienced alongside other advertisements that pull consumers in opposite directions. In short, patterns of environmental behavior suggest an intriguing paradox that is not easily resolved. An answer is sought here by reference to four possible explanations—self-interest, political ideology, personal efficacy, and product advertising.

Self-Interest

One potential reason for the high frequency of consumer behavior measured in recent surveys is simple. Not all green activity in the marketplace can (or should) be considered public-regarding. While individuals can undertake many different activities that benefit the natural environment, those actions may be motivated by a host of factors, including self-interest, financial and otherwise.[21] For example, a 2000 survey conducted by the Gallup Organization asked respondents which activities they had participated in "in the past year." As chapter 2 noted, while most reported that they had recycled glass, plastic, and aluminum cans

(90 percent), recycling is now mandatory in many cities and municipalities, often with curbside pick-up.[22] As some scholars point out, activities such as these are a form of "forced" behavior change.[23] In other states, such as Maine and Vermont, consumers are reimbursed a small deposit fee after bottles are emptied and returned to community redemption centers. While some families, of course, recycle household waste out of a genuine concern for the environment or an intrinsic sense of personal satisfaction,[24] others undoubtedly do so because of a financial incentive or a need to comply with local ordinances.[25] The latter are factors that clearly question whether recycling should be seen as an honest reflection of the public's commitment to environmental protection.

On other occasions, too, the Gallup Organization probed its respondents about whether they had improved their home's insulation or heating and air conditioning systems, or whether they cut water use, drove a more fuel-efficient car, or carpooled with others. The motivation of respondents in performing these tasks is equally unclear. As efforts that conserve energy and natural resources, each could be considered environmentally friendly, and yet the financial benefits received by reducing individual energy and transportation costs are not trivial. In short, even though some scholars view environmental activities like recycling[26] and energy conservation[27] as altruistic and public-regarding, that need not be the case. When it pays to buy green, the influence of self-interest cannot be overlooked.

Finally, while self-interest is often narrowly defined in monetary terms, environmental issues relating to the protection of public health and safety suggest the need for a more inclusive definition. Mark Baldassare and Cheryl Katz find that respondents who worry about the impact of environmental problems on their own personal well-being are more likely to recycle, conserve water, buy environmentally safe products, and limit their driving to reduce air pollution—an effect not matched in predictive ability by any other social, political, or demographic trait.[28] Another study suggests that choosing to purchase organically grown fruits and vegetables in the supermarket represents an effort on the part of some consumers to protect their families against a perceived cancer risk associated with exposure to pesticides.[29] Both of these cases suggest that certain environmentally responsible activities should be seen as a variant reflection of self-interest.

In the end, each of these issues suggests a hypothesis that can be tested empirically using survey data. First, if this set of explanations holds, the frequency of environmental behavior should be higher when self-interest, financial or otherwise, is apparent. Second, if green consumerism is motivated by self-interest, respondents also should be price conscious. Willingness to pay a premium price for environmentally safe consumer goods should be rather weak. Moreover, when compared to political activities like voting, socioeconomic traits such as income should have a greater impact on consumer behavior, where the incentive to consider narrow self-interest is at its strongest. Finally, if consumers are swayed by self-interest rather than the public interest, market choice should be more strongly motivated by environmental concerns close to home rather than by a general concern for the environment as a whole.[30]

Political Ideology

Extensive research on the social bases of environmental concern suggests a second possible reason for the popularity of green consumerism in surveys. Scholars have long shown that certain environmental attitudes are associated with a liberal political ideology—a point reinforced by empirical results presented in chapter 4 of this book.[31] Environmental regulation is commonly opposed by conservative ideologues because of its financial cost and the extensive government interference in the market economy it requires. Many voters, too, are cautious about expansions of federal and state bureaucracy that oversee private land use decisions or create additional layers of "red tape." For these reasons, in the political realm we would fully expect environmental attitudes and behaviors to be closely associated with standard measures of liberalism and conservatism.

In stark contrast, however, the marketplace imposes no such barriers. In fact, buying green is entirely consistent with the laissez-faire or market-based approach to environmental protection favored by conservatives. In this way, environmental marketing appeals to a wide audience. Environmentally concerned individuals who are unwilling to cross party lines to vote for green political candidates or strict regulatory policies might well find the marketplace a more ideologically compatible outlet for environmental expression. If this reasoning holds, data should show that ideological cleavages are relatively weak in the area of consumer behavior, especially when compared to activities that are distinctly political in nature.

Personal Efficacy

James Buchanan's seminal essay on individual choice in the voting booth and the marketplace discusses a third possible reason for the popularity of green consumerism.[32] Environmentally friendly products may appeal to consumers because their purchase allows greater certainty, responsibility, and control over the outcome. Unlike voters who have only a single vote to cast, consumers have highly divisible incomes. They are free to select the combination of goods and services closest to their own environmental views and (under ideal conditions) can be assured of the direct and immediate results of their actions.[33]

Based on their experiences in the marketplace, therefore, we might expect consumers to be empowered by a stronger sense of personal efficacy. This hypothesis has received support in the literature, albeit indirectly. While some scholars continue to insist that Americans buy green in direct proportion to their income and education levels,[34] many studies attempting to identify environmentally conscious consumers find that psychological variables, such as perceptions of efficacy, have far greater explanatory value.[35] If this logic bears out, feelings of personal efficacy should correlate less strongly with market behavior when compared to those things that depend upon collective action for their success (such as joining an environmental group or contributing money to an environmental cause).

Product Advertising

Studies of consumer response to advertising suggest a fourth and final explanation to our original paradox of the popularity of green consumerism and the lack of green success on election day. Consumers in the marketplace may be willing to purchase environmentally friendly products because of successful marketing efforts that draw their attention in that direction, through the use of product labeling or even green shelf tags.[36] Evidence here is impressive. Experimental research has found that environmental labeling is "significantly more persuasive" than more traditional appeals to cost savings.[37] Others likewise argue that under experimental conditions, the use of an altruistic frame increased respondents' vote intentions for a container deposit law more than did an appeal to narrow self-interest.[38] In examining willingness-to-pay (WTP) for a variety of consumer goods, another team of scholars finds that the use of symbolic framing (such as appeals to environmental concern or patriotism) decreases price sensitivity and increases the variance of the WTP function, compared to the presence of instrumental cues alone.[39] Given that symbolic appeals of this nature often outperform instrumental considerations at higher product prices, they contend that environmentally responsible products may ultimately succeed in the marketplace, even with a premium price tag, if the purchase of those products can be cognitively linked though advertising to an appealing goal, such as saving dolphins from tuna nets or preserving a tropical rain forest.

This line of work suggests one final hypothesis—that if consumers buy green because of the persuasive nature of environmental marketing, labels that tout products as being environmentally friendly in one way or another should provide a powerful influence on consumer choice, largely independent of whatever latent environmental concern individuals bring with them into the marketplace.

Data Analysis

The explanatory power of each of the factors outlined above is explored in this chapter, albeit in a somewhat piecemeal fashion, by employing a variety of datasets for a variety of different purposes. It is an approach born of necessity. Comparing political and economic behavior on environmental issues directly is a difficult task. Because of differences in purpose and scope, few surveys include measures of both kind. High-quality academic sources of data such as the National Election Study (NES), have turned their attention to environmental issues only recently and tend to focus exclusively on measures of political interest. In contrast, private polling firms hired to administer surveys for business clients often probe their respondents thoroughly about consumer behavior but ask for very little information about political activities or partisan viewpoints. Moreover, because private data are gathered for financial gain, they are seldom made available to an academic audience for analysis.

Given that no single survey adequately addresses all of the hypotheses raised in this chapter, three different polls are used here—an April 2000 Gallup poll and a 1993 study conducted by Cambridge Reports Research International (both cited earlier), as well as the 1994 NORC General Social Survey (GSS). All provide insights into the motivations that underlie environmental action.

The Influence of Self-Interest

Recall that one possible explanation for the popularity of green consumerism was a simple variant of self-interest. Perhaps Americans take part in some environmental activities, such as energy conservation, out of a desire to reduce their own monthly utility bills. That suspicion is supported in looking at any number of recent polls and surveys.[40] For reasons easily explained by appeals to self-interest, Americans tend to prefer activities that conserve natural resources while at the same time save money and expense.

In the April 2000 poll conducted by Gallup cited in chapter 2, respondents were asked which activities they or other members of their household had done "in recent years to try to improve the quality of the environment." Few reported that they had contacted a public official about the environment (18 percent) or been active in an environmental organization (15 percent), both activities that are clearly political in nature. Even among small numbers, unless those actions are grassroots in nature (that is, directed at a problem of local environmental concern, such as the siting of a hazardous waste facility in one's own backyard), self-interest probably provides a minimal incentive.

In contrast, large majorities in the Gallup study cited their conservation of energy (83 percent) and water (83 percent). Most reported that they had bought "some product specifically because [they] thought it was better for the environment than competing products" (73 percent). In the end, while the overriding motivation that pushes individuals into action may be uncertain here, the pattern of responses across activities is telling.

With few exceptions, actions that contain elements of financial self-interest are more commonly reported than those that do not.[41]

If green consumerism is indeed private rather than public regarding, respondents also should be price conscious. Aside from issues like energy efficiency and water conservation, which tend to reduce monthly household expenses, willingness to pay a premium price for environmentally friendly products that offer no tangible financial interest in return should be weak. Those expectations, however, are only partly confirmed using data from a 1993 Cambridge Reports Research International poll. Unlike the April 2000 Gallup survey that questioned respondents in a broad way about their participation in certain environmental activities "in the past year" (where answers were coded into a simple yes/no dichotomy), the Cambridge poll probed a similar national sample of Americans on the frequency of their behavior, with a particular emphasis on purchases made in the marketplace. While narrower in its focus than the Gallup survey, the Cambridge results are significant in that respondents were pressed to consider the added cost of environmentally safe goods and services. That unique battery of questions is particularly useful here.

As table 8.1 shows, most respondents in the Cambridge poll seemed willing to pay a higher price for environmentally friendly products. Indeed, the distribution of willingness-to-pay shows little variance across items that range from automobiles and gasoline to detergents and paper products, with most willing to pay at least 5 percent more. Consumer support erodes quickly at levels exceeding 5 percent, but the willingness of many to pay even a small amount more, despite the thrifty tendencies of the marketplace, suggests that consumer goods may indeed be less price sensitive when framed in environmental terms.[42]

Even more remarkable than a willingness to pay for environmental goods, however, is the extent to which those considerations transcend typical social and demographic cleavages. While green consumer activity is more frequent among women, the young, and the well educated, the effect of household income in two regression equations presented in table 8.2 is statistically indistinguishable from zero. Quite simply, income fails to explain either the frequency of green consumer purchases or the willingness of some to pay a higher price for those products. Overall, too, all of these social and demographic factors (which include political ideol-

Table 8.1
Willingness to Pay for Environmental Products

Now I am going to read you a list of products. Please tell me how much more you would be willing to pay for each product if it were environmentally friendly. Would you be willing to pay 5 percent more, 10 percent more, 20 percent more, or at least 30 percent more? If you wouldn't be willing to pay anything more for the product if it were environmentally friendly, please just say so.

Product description	No more	5%	10%	20%	At least 30% more	Mean	Standard deviation
Paper products made out of recycled paper	23.0%	36.7%	22.0%	10.2%	8.1%	1.45	1.18
Household products such as kitchen and bathroom cleaners	23.8	37.1	23.2	8.8	7.2	1.38	1.15
Garden products such as insecticides and fertilizers	33.2	26.9	22.4	9.7	7.9	1.32	1.25
Plastic packaging or containers made of recycled plastic materials	25.5	39.1	20.6	9.5	5.3	1.30	1.11
Detergents	26.4	38.3	21.2	8.9	5.2	1.28	1.11
Automobile	37.5	28.0	21.2	7.2	7.9	1.24	1.23
Plastic packages or containers made with less plastic	27.8	41.1	19.4	7.6	4.1	1.19	1.05
Gasoline	33.7	34.2	20.5	6.0	5.6	1.15	1.12

Source: Cambridge Reports Research International (July 15–27, 1993).

Table 8.2
Determinants of Green Consumer Behavior

Independent variables	Model 1 Frequency of buying green products		Model 2 Willingness to pay more for green products	
	OLS slope estimate	Standard error	OLS slope estimate	Standard error
Age	−0.37	0.26	−1.23***	0.17
Education	0.11	0.25	0.42**	0.16
Income	−0.09	0.19	−0.03	0.12
Race	0.52	1.20	−0.38	0.77
Gender	−4.65***	0.75	−0.54	0.47
Political ideology	0.84*	0.41	1.29***	0.26
Intercept	36.09	1.72	10.96	1.07
Mean	34.04		10.58	
Number of cases	909		888	
R-square	0.050		0.103	

Source: Cambridge Reports Research International (July 15–27, 1993).
Notes: See appendix for question wording.
* $p < .05$.
** $p < .01$.
*** $p < .001$.

ogy) explain only a tiny fraction of the variance observed in either dependent variable.

In the end, the limited effect of primitive self-interest on environmental activities in the Cambridge study is largely consistent with Robert Rohrschneider's findings cross-nationally. He argues that "self-interest concerns are an important force in the economic realm of politics, but the emergence of non-economic issues in advanced industrialized nations may diminish the explanatory value of self-interest motives."[43]

Do those same limits apply, however, when self-interest is more broadly conceived? Non-economic issues like environmental quality may suggest not that the explanatory value of self-interest has declined in a post-materialist era but rather that its impact can and should be measured in new ways. The NORC General Social Survey (GSS) offers an intriguing and unique test of this idea. In 1993 and again in 1994, the GSS questionnaire included a lengthy battery of environmental items, including six well-paired measures that asked respondents to rate how dangerous certain environmental problems were to the environment "in general" and to "you and your family." The general dimension tapped by the first question is similar to an emphasis on "sociotropic" concern,[44] while the second more specific measure relating environmental quality to a respondent's personal well-being seems consistent with a broader notion of self-interest.[45]

Which target of environmental concern motivates consumer behavior —in this case, the frequency with which respondents make a "special effort" to buy pesticide-free fruits and vegetables? GSS data from 1994 initially show little difference. When asked to rate the danger of pesticides and chemicals used in agriculture, most respondents gave similar answers to both questions. In fact, the degree of correlation between the general and specific measures was so high (with a Pearson's r of .86) that both could not be included as independent variables in the same regression equation without causing of a problem known as *multicollinearity*.[46] Instead, a new variable was created by subtracting a respondent's general perception of danger from pesticide use from its more personal component, producing a single scale ranging in value from -4 to $+4$.

After controlling for household income (which might well influence an ability to buy organic produce at higher cost), regression results for this scale (see table 8.3) show that a higher relative degree of personal concern for the dangers of pesticide use does in fact influence the purchase of pesticide-free fruits and vegetables, but even in this case the influence of self-interest is surprisingly small.

The Impact of Political Ideology

If self-interest is of limited value in understanding the appeal of environmental goods and services, perhaps explanations that focus on the unique character of the marketplace fare better. While the environmental battery used by the NORC General Social Survey in 1994 concentrated on collective action, asking if respondents had signed a petition about an environmental issue or given money to an environmental cause, the Cambridge survey administered in 1993 focused almost exclusively on consumer choice. As noted earlier, the Cambridge poll also added one comparable measure of political behavior, asking respondents how often

Table 8.3
Personal versus General Concern for the Environment as a Predictor of Consumer Behavior

And how often do you make a special effort to buy fruits and vegetables grown without pesticides or chemicals?

Independent variable	Probit estimate	Standard error
Difference in perceptions of danger from pesticides to "you and your family" (+) and the environment "in general" (−)	0.15*	0.07
Household income	−0.02**	0.01
Intercept 1	−1.08	
Intercept 2	0.73	
Intercept 3	1.63	

Source: NORC General Social Survey (1994).
Notes: The primary independent variable of interest used in this model is calculated as the numerical difference between responses to two measures. The first asks "*In general*, do you think that pesticides and chemicals used in farming are [extremely (5), very (4), somewhat (3), not very (2), not dangerous at all (1)]?" The second question reads: "Do you think that pesticides and chemicals used in farming are [extremely (5), very (4), somewhat (3), not very (2), not dangerous at all (1)] *for you and your family?*" Since these two measures are so highly correlated, both could not be included in the same regression equation without the risk of multicollinearity. Instead, general perceptions of environmental danger were subtracted from personal concern to create a scale that ranges in value from −4 to +4. Mean perception of danger to "you and your family" = 3.31. Mean perception of danger to the environment "in general" = 3.35. Pearson's r (personal, general) = 0.86
*$p < .05$.
**$p < .01$.
***$p < .001$.

they "vote for a political candidate primarily because the candidate took strong environmental positions." Part of the same group of questions that asked how often they "buy a product because the label or advertising said it was environmentally safe or biodegradable," both measures used here share an identical response format ranging from a low to high frequency of activity.

Given that, we would expect political ideology to be more predictive of vote choice, and indeed that is the case. As regression models in table

Table 8.4
Determinants of Environmental Behavior

Now I am going to read a list of things people have done because of their concern for the environment. Please use a scale of "1" to "7," where "1" means "never have done that thing" and "7" means "very frequently do that thing" to tell me how often you do each thing I read:

Independent variable	Model 1 Vote for a political candidate primarily because the candidate took strong environmental positions		Model 2 Buy a product because the label or advertising said it was environmentally safe or biodegradable	
	OLS slope coefficient	Standard error	OLS slope coefficient	Standard error
Age	−0.01	0.05	−0.05	0.04
Education	0.04	0.05	0.00	0.04
Income	−0.03	0.03	0.02	0.03
Political ideology	0.25***	0.07	0.08	0.07
Race	0.24	0.22	0.13	0.20
Gender	−0.33*	0.13	−0.64***	0.13
Intercept	3.20	0.31	4.95	0.29
Mean	3.41		4.70	
R-square	0.022		0.028	
Number of cases	1,001		1,014	

Source: Cambridge Reports Research International (July 15–27, 1993).
Notes: See appendix for question wording.
* $p < .05$.
** $p < .01$.
*** $p < .001$.

8.4 show, the effect of ideology on voting green is both statistically significant and substantively strong, while its imprint on buying decisions is statistically indistinguishable from zero. Consumer choice, in this sense, does seem to transcend ideology and in doing so allows environmental products to appeal to a wider audience.[47]

Certainty and Control over Decision-Making

As James Buchanan and Gordon Tullock note, the marketplace is unique in other respects as well. An individual participant in collective choice is but one of many decision makers.[48] Contributing money to an environmental group offers no guarantee that others will do likewise to a degree that is sufficient to fund a successful media campaign or a lobbying effort. There is no longer, as they note, a "one-to-one correspondence between individual choice and the final action."[49] In contrast, control over (and responsibility for) choices made in the marketplace rest squarely on the shoulders of the individual. Do considerations such as these help us to understand why individual choice continues to take precedence over collective action, even on large, seemingly intractable issues like the environment, where the ability to effect change at the individual level seems slim? A return to the 1994 NORC General Social Survey offers a tentative answer.

The GSS asked respondents about a variety of environmental activities. Some of these questions focused on involvement in collective tasks—joining an environmental group, signing an environmental petition, contributing money to an environmental cause, and so on—while others were targeted to individual behaviors, such as recycling, buying pesticide-free produce, and cutting back on the use of a car. The GSS also asked respondents about the extent to which they agreed with the following statement: "It is just too difficult for someone like me to do much about the environment." Degrees of agreement on this item were scored high, while degrees of disagreement were scored low.

If the marketplace provides consumers with a greater degree of rationality and control, feelings of personal efficacy should associate less strongly with individual behavior, at least when compared to activities that depend on the active participation of others for their success. Correlations reported in table 8.5 generally support that expectation. Joining an environmental group, contributing money to an environmental cause, taking part in an environmental protest, signing a petition about an environmental issue—all clearly diminish when efficacy is low, with correlation coefficients ranging from $-.33$ to $-.48$. The relationship between efficacy and individual action is far weaker in comparison, ranging from just $-.04$ to $-.20$. In short, while scholars have long suspected that low perceptions of efficacy are an "effective deterrent" to environmental behavior,[50] evidence here suggests that actions taken in the marketplace might help Americans to feel better about their own ability to effect environmental change.

Table 8.5
Correlations Between Environmental Behavior and Personal Efficacy

	Personal Efficacy "It is just too difficult for someone like me to do much about the environment"
Collective behavior:	
Member of environmental group	−0.33
Signed environmental petition	−0.46
Contributed money to environmental group	−0.48
Taken part in environmental protest	−0.40
Additive scale of collective behavior items	−0.43
Individual behavior:	
Frequency of recycling	−0.20
Frequency of buying fruits and vegetables grown without pesticides and chemicals	−0.04
Cut back on driving car for environmental reasons	−0.20
Additive scale of individual behavior items	−0.17

Source: General Social Survey (1994).
Note: All measures of association reported above are gammas.

The Power of Environmental Advertising

One final difference between market choice and social choice remains. As Donald Philip Green notes, political campaigns "tend to be two-sided affairs; and the potency of negative symbols such as bureaucratic waste, big government, and undeserving beneficiaries" can undermine support for even the worthiest of causes. Product advertising is seldom so contentious.[51] Labels that promote a product as being environmentally friendly have to compete for attention alongside other powerful factors, including appeals to cost savings and brand loyalty, but rarely is information available at time of decision to contest a company's environmental claims. In other words, because the marketplace lacks the balance provided by reciprocal debate, we should expect environmental marketing to have a stronger impact on consumer choice.

While that underlying logic is supported by a variety of studies that examine the power of advertising and of environmental marketing in particular,[52] it is a difficult premise to address here given the limits of available survey data. The April 2000 Gallup survey on environmental attitudes does not address factors such as advertising that should be a close, proximate influence on market choice. Likewise, since the 1994 NORC General Social Survey contains only a single measure of consumer behavior, little effort is made to better understand the conditions under which those purchases are made.

The Cambridge Reports Research International poll conducted in July 1993, however, contains a reasonable start. Respondents were asked if "in just the last week" they had "really read the label on a product to find out whether or not it is better for the environment." They were also asked to rate how important environmental labels were to them when making purchase decisions on household or garden products. The interaction of these two variables on the frequency of green consumer behavior is plotted visually in figure 8.2. Important here are the differences in the slope of each regression line, which demonstrate that reading labels to see if a product is good for the environment seems to work more effectively on consumers who do *not* consider environmental attributes to be particularly important.

Although counterintuitive at first glance, that result, suggests that environmental marketing—or at least the environmental framing used in surveys—is exceptionally powerful in the way that it alters the market preferences of those least committed to the environment. For consumers who consider environmental attributes to be "absolutely essential," labels are likely to provide information but little more. Whatever persuasion or motivation might be needed to effect a green purchase comes from the consumer directly and the level of environmental concern they bring with them into the marketplace. But for those less convinced of the value of environmental products, that information might well provide a new and important source of differentiation, subtly directing attention and increasing the cognitive availability of certain considerations at the expense of others, such as cost.

Conclusions

With a growing number of businesses jumping on the environmental bandwagon, green consumerism has emerged as a major social and eco-

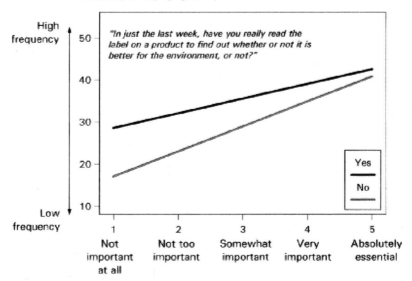

Frequency of buying "green" products

High
frequency

"In just the last week, have you really read the
label on a product to find out whether or not it is
better for the environment, or not?"

Yes

No

Not
important
at all

Not too
important

Somewhat
important

Very
important

Absolutely
essential

"When you shop for household and garden products,
how important is it that the products you buy are
labeled environmentally friendly in some way or
another?"

Figure 8.2
The Impact of Environmental Labels on Buying Green
Source: Cambridge Reports Research International (July 15–27, 1993).
Note: The dependent variable used in this chart is an additive scale created from
eight measures of green consumer behavior. Respondents in this battery were
asked how frequently they do each of a variety of activities on a seven-point
scale, where 7 indicates that they "very frequently do that thing." Coded answers
to these questions were added to create a scale ranging in value from 8 to 56.

nomic force over the past ten years, allowing environmental issues to
expand far beyond the sphere of politics into the daily lives and lifestyles
of the American consumer. Yet while we might view environmental
choices made in the marketplace as an extension of political attitudes
and beliefs, scholars have long cautioned that the actions of voters and
consumers are likely to diverge as conflicting emotions of self-interest
and altruism take their place. The patterns of behavior delineated here,
however, suggest that traditional expectations are at times false or at the

very least misplaced when it comes to the environment. Cross-pressured in many different ways, James Buchanan and Gordon Tullock conclude simply that "man is, indeed, a paradoxical animal."[53] The data presented in this chapter offer a preliminary step in explaining how and why.

First, consumer behavior seems to offer fewer barriers and richer rewards. Loyal Republicans may not be willing to break party ties to vote for green political candidates or strict regulatory policies, but choices made in the marketplace are largely shielded from those effects. Moreover, a market-based approach to environmental regulation is entirely consistent with conservative ideology. By allowing consumers to look beyond government regulation for solutions to complex environmental problems on issues such as air pollution and solid waste disposal, environmental products provide Republicans with an alternative way to express genuine environmental concern.

As James Buchanan notes, too, choices made in the marketplace seem to offer citizens a greater degree of certainty and rationality. With a highly divisible source of income, consumers can select the combination of goods and services closest to their own environmental views and can be assured of the direct and immediate result of their actions. In contrast, voters on election day often face a limited number of alternatives, and given majority rule those casting ballots can never be certain that their votes will lead to a preferred result. Political pressures may encourage citizens to weigh the public interest, but the marketplace rewards consumers directly with a palpable sense of efficiency and control.

Finally, the body of evidence offered in this chapter suggests that the manner in which environmental problems are framed by advertisers and political opponents may give rise to a different ordering of preferences, influencing the value that both voters and consumers place on environmental goals relative to other social, economic, and political priorities. In other words, the way in which environmental issues shape behavior may be dependent ultimately on the cues or symbols that are cognitively linked to it.[54] Environmental marketing would seem to succeed by investing private goods with public meaning, by distracting consumer attention away from considerations of cost, and by exploiting the influence of the public-minded citizen. While environmental quality is tra-ditionally viewed as the quintessential public good, those problems in many ways span the chasm between public and private, blurring lines of distinction between citizen and consumer in the process. Given Buchanan and Tullock's gentle reminder that "the same individual participates in both processes,"[55] the fluidity of those boundaries suggests that public and private decision makers might fare better in the long run by focusing on the psychology of choice rather than its location.

Conclusion
Rethinking Environmentalism

Perhaps the greatest problem the environmental movement will have to face lies not in the external world but inside the heads of the American people.
—Philip Shabecoff[1]

In 1992, in reflecting on the popular appeal of environmental issues during the course of a presidential campaign that would bring a Democrat to office for the first time in twelve years, along with an avowed environmentalist as his running mate, Philip Shabecoff wrote that "virtually every aspect of our personal lives, from the food we eat to the packages we use, to the way we drive and the fears we have for our children's future, has been altered by environmentalism."[2] Those concerns, of course, have melded into a powerful social movement whose importance ranks alongside the peace, labor, women's, civil, and human rights movements of the past century.[3] With a dedicated cadre of political activists advancing environmental interests in Washington and abroad, environmental values have led to massive changes in business, science, education, and law.

Yet in a field increasingly dominated by expertise and bureaucracy, public opinion—the grassroots of this green revolution—is potentially no less significant in understanding the path of environmental progress.[4] While the legacy of the environmental movement will be measured surely by its legislative achievements, so too will it surely be judged by its ability to persuade average Americans to back up their words with action and to change their voting patterns, buying habits, and lifestyles. To explore that topic in depth, as this book has done, is to speak to the meaning and relevance of environmentalism in American politics today.

The purpose of this final chapter is to weave together disparate pieces of that debate and to draw out its logical consequences, revisiting several broad themes and questions posed at the outset along the way.

A Poverty of Language

In writing for *Smithsonian* magazine, John Wiley struck a resonant chord when he said that "all of us are environmentalists." There is a shared reverence for life, he argued, and a love of place that unites "river keepers, protesters in treetops, middle-aged birders, children discovering aquatic invertebrates, tropical biologists, environmental lawyers and toddlers stumbling after butterflies."[5] On its face, of course, to believe as much may seem instinctive (even obvious), for as Wiley is quick to point out, we would all "just as soon have clean air and water" than not. But an important message underlies that simple truth nevertheless. Today, most Americans accept environmental protection as a legitimate cause for individuals, corporations, and governments alike, and the significance of that evolution in attitudes should not be underestimated.[6] We no longer debate *whether* to protect the environment but rather where, when, how, under what conditions, and at what expense. The complexity of those new decision rules, however, forces us to wonder, as Wiley did, about the meaning (and overuse) of the labels we attach. To apply the term *environmentalist* equally to all is, in his words, "to embrace a poverty of language" badly in need of revision.[7]

Based on the multitude of polling data presented in this book, Americans tend to view environmental issues in complex ways not easily explained using vague and imprecise vocabulary. For instance, to describe an environmentalist merely as someone who is concerned with the quality of the natural environment (as many standard dictionaries and encyclopedias do) seems unhelpful, even tautological. A cultural consensus formed around that singular preference may indeed be a welcome achievement, but it is unlikely to provide much insight or nuance into the true nature of public thinking on the environment. To respond fully to the latter requires that close attention be paid to a broader list of qualities, ones that collectively paint a complicated and at times contradictory picture.

For example, while public concern for the environment is strikingly high, it is neither strong nor salient. Americans place genuine value and priority on environmental quality, to be sure, but they also support lowering crime rates, improving public education, and maintaining a strong economy (among other things), all of which easily surpass the environment in polls. Data likewise suggest that voters and taxpayers are sensitive to the costs associated with protective environmental policies, often preferring a moderate course of action when faced with scientific uncertainty or when painful choices are forced between the environment and other equally desirable goals. For these reasons, surveys that measure commitment to the environment by reference to the environment alone, without regard to competing priorities and budgetary constraints, are at best unrealistic and at worst misleading. Public interest in the environment is sincere and well intentioned, but as with any issue of social or economic importance, it is not without limits that constrain political opportunity.

Second, while environmental concern is widespread, behavior is not. The diversity of issues that now fall under the umbrella of the environmental movement (which range from toxic waste disposal to wilderness and wildlife protection) create strength in numbers by helping environmentalists seam together an agenda that appeals to people from all walks of life, with all manner of interest in nature. At the same time, however, members of this "sympathetic public" often fail to take relatively simple and inexpensive steps toward environmental action.[8] While many households do recycle and some purchase green products in the marketplace, evidence suggests those decisions are more likely when based on tangible incentives or when the need for personal sacrifice is slight. Quite simply, whether rich or poor, black or white, liberal or conservative, Americans are united in wanting something to be done about the environment, but they remain hesitant across the board to include themselves as a part of that solution.

Third, while the environment has enjoyed remarkable staying power on the national political agenda over the past thirty years, public commitment to those issues can be somewhat fickle, moving in cycles that visibly advance and retreat over time. The fact that progress in some areas has been quickly offset by the emergence of other pressing concerns, including in recent years ozone depletion, global warming, and urban sprawl, has allowed the environmental movement to reinvent itself and rejuvenate its support time and time again.[9] But in the end, the willingness of voters and taxpayers to fund policies designed to address those problems depends not on anxiety and evidence alone but rather on a number of unpredictable factors, including media attention and the health of the national economy.[10] Both allow moments of opportunity but also enforce considerable constraint on those who lobby on behalf of stronger regulations.

Finally, public attitudes on the environment are surprisingly consistent across a wide range of survey instruments. Under the best of circumstances, this consistency could well mark the development of a new social paradigm or belief system capable of reordering the relationship between man and nature. But that assumption is marred for two reasons. The close entwining of environmental attitudes with economic goals and conditions suggest that Americans remain firmly committed to both, including a market economy dedicated to the pursuit of material comfort and abundance. Moreover, public opinion at key moments seems unsupported by factual knowledge, particularly on issues like global warming that require a degree of technical or scientific judgment.[11] Since answers to questions on environmental topics can be shaped by social desirability and even altered by the way in which questions are phrased in surveys, the content and sophistication any new environmental belief system remains in doubt.

Barriers to Action

Based on attitudes that fall into anomalous patterns, it is not surprising that widespread environmental concern has failed to translate into greater public resolve, especially in the voting booth. As Willett Kempton, James Boster, and Jennifer Hartley write, the "usual conclusion drawn from modest environmental actions is that modest actions indicate a modest commitment."[12] Yet several scholars, including Kempton and his colleagues, stress instead the existence of certain "barriers to action."[13] When viewed expansively, it is an theory that can give helpful insight into the body of evidence described above by proposing a simple explanation—that supportive attitudes are a necessary but hardly sufficient condition for environmentally responsible behavior.

One barrier described in this book may be feelings of personal inefficacy fueled by the sheer enormity (and apparent intractability) of environmental problems. "After all," as Steven Teles writes, "despite the appeal of environmentalists to 'think globally, act locally,' the former injunction typically overwhelms the latter."[15] The chronic low salience of environmental issues represents a second potential obstacle. It is evident that most Americans are concerned about the environment, and yet those issues rarely generate the power and immediacy needed to push into the top tier of voting preferences. Without a strong sense of effectiveness or the cognitive awareness to ignite it, thoughts and words lose momentum in the causal chain connecting attitudes to behavior.

Inadequate information represents a third possible impediment to effective environmental action. According to an annual report card developed by the National Environmental Education and Training Foundation (NEETF) in Washington, D.C. (in conjunction with the Roper Organization), many Americans receive a failing grade in basic environmental knowledge and persist in perpetuating certain myths and misconceptions.[16] For instance, a surprising number of those polled for NEETF's 1998 study thought that ozone-depleting chlorofluorocarbons came primarily from aerosol spray cans (rather than refrigerators and air conditioning units), despite a ban on that particular use in 1978. A majority of respondents also believed that the major benefit of recycling paper products was to save trees rather than to reduce the volume of solid waste sent to landfills by consumers. Moreover, just 16 percent knew that the major source of petroleum pollution in rivers, lakes, and bays is runoff from residential use rather than spills caused by oil rigs, tankers, and refineries.[17] In each of these cases, more accurate knowledge might press average citizens to personalize environmental risk in ways that encourage individual responsibility and involvement.[18]

Finally, as rational choice theorists are quick to point out, environmental behavior is often blocked by collective action problems that allow free riders to enjoy the benefits of environmental protection unencumbered by cost.[19] Under those circumstances the importance of finding tangible incentives to motivate behavior should not be underestimated.

According to Matt Ridley and Bobbi Low, "the environmental movement has set itself an unnecessary obstacle by largely ignoring the fact
that human beings are motivated by self-interest rather than collective interests,"[20] a point well illustrated by Garrett Hardin's cautionary
tale—"The Tragedy of the Commons."[21] Others suspect, too, that participation in environmentally responsible activities will remain low regardless of awareness and concern, unless the environment can be linked
to social arrangements, such as curbside recycling programs, that "help
reduce the costs of compliance and facilitate cooperative action."[22]

Practical Advice

The opening chapter of this book urged that a clear understanding of
public opinion on the environment was necessary for a variety of reasons. Given the empirical efforts made in that direction in the intervening
pages, it seems logical to list the lessons learned, and to attempt to
translate academic evidence into practical advice. One possible response
to that challenge is purely descriptive—to outline in some detail the
boundaries of popular support and the constraints (and opportunities) it
imposes on lobbyists and lawmakers alike. But a second is more pro-
scriptive. It is to insist that energy and attention be committed to exploring how the "barriers to action" described above might be lowered
in careful and creative ways. The sum of both of those approaches is
detailed next.

Building a Civic Environmentalism
When Gregg Easterbrook published a lengthy book titled *A Moment on
the Earth* in 1995, he was greeted by a hailstorm of criticism from within
the environmental community. Widely denounced by activists in newspapers, press releases and radio debates, his plea for optimism in light of
what he considered certain environmental progress, soon caused him
to become, according to one sympathetic reviewer, "Green Enemy
No. 1."[23] While some attacked his work directly, arguing that it was
"replete with errors and misinterpretations of the scientific evidence,"[24]
others took issue above all with his hopeful tone, insisting that his "dec-

laration of victory" over environmental degradation was dangerously premature.[25]

The controversy surrounding Easterbrook is an interesting one. As one newspaper columnist bluntly put it on the eve of Earth Day 2000, "Environmentalists have one of the great American success stories to tell, if only they would tell it."[26] Yet in the competition for public attention, Easterbrook argues, a rather "peculiar intellectual inversion has occurred in which good news about the environment is treated as something that ought to be hushed over, while bad news is viewed with relief."[27] In fact, he insists that environmentalists fear signs of success in the belief that it might "dilute the public sense of anxiety" that serves for them such a useful political purpose.[28] If a strategy of unrelenting environmental pessimism is misguided, as Easterbrook believes it is, how should environmentalists recapture in mass the excitement and exhilaration of their cause? Collectively, the results of this book speak to the difficulty of that task and to the importance of distinguishing rhetoric that leads alternately to what John Immerwahr calls stagnation and engagement.[29]

One possible response to poll results, says Immerwahr, is to insist "that the public is apathetic and needs to be awakened to the seriousness of the consequences of ignoring" environmental problems. Consistent with that understanding, environmentalists might insist on a "concerted educational effort to convince the public of the importance of these issues." But, he says, to do so represents a fundamental miscalculation. After all, evidence from national surveys suggest that Americans *are* concerned about environmental degradation and are persuaded of its dangers. It is simply that other values compete for scarce energy and attention, a blunt reality that is at once understandable and stubbornly pragmatic. Immerwahr warns that exaggerating the seriousness of environmental problems to a lay public with the intent of intensifying concern might instead "increase frustration and apathy rather than build support," encouraging Americans to believe that those problems are intractable.[30] "[W]hat the public is most skeptical about," he says, "is not the existence of problems but our ability to solve them."[31] Others, too, warn that environmental rhetoric is likely to fail when "couched in terms of sacrifice, selflessness, or, increasingly, moral shame."[32]

If getting the message right is paramount for environmentalists intent on political persuasion, what strategies should they employ? What language encourages popular engagement and forward momentum? One possible model lies in recent attempts to define a new "civic environmentalism," one that promotes goals that are less elitist and more hopeful, focused as much on the pride of ordinary places as on pristine wilderness. It is, as William Shutkin notes, a movement "that ultimately finds a comfortable place for people in nature" by focusing on healthy and sustainable communities that make our environment one worth living in.[33] As he writes, "the best kind of American environmentalism" recognizes that the most viable solutions are often those "inextricably linked to social, political, and economic issues," in essence to "community life in its totality."[34]

Political Obstacles and Opportunities

To recognize Shutkin's sense of community is also, of course, to acknowledge that environmentalists must operate within existing political realities, needing at times to embrace compromise over ideological purity in order to get things done. In 1990, California voters overwhelmingly rejected Proposition 128, a ballot initiative more suitably known as "Big Green." While well intentioned, this measure planned to merge numerous environmental issues under one enormous banner. Its goals were ambitious and complex—among them to ban the use of nearly twenty cancer-causing pesticides, to prohibit new oil and natural gas drilling along the California coast, to limit emissions of greenhouse gases, to outlaw the clear cutting of old-growth redwood forests, and to create a new elected post of "environmental advocate" to head enforcement of the law. As accumulated experience on environmental ballot propositions suggests, however, legislative success is more likely when taken in small, incremental steps that avoid uniting political enemies.

Politicians seeking elective office are constrained in similar ways by the narrow range of acceptable positions on the environment they can take, and by a still narrower audience of uncommitted voters. To use the bully pulpit to heighten a sense of issue salience from the top down should reward candidates who promote policy differences with clarity and aggressiveness, especially in state and local campaigns where environmental issues enjoy less competition for room on a crowded political agenda. But as James Q. Wilson observes, "It is hard ... to make a campaign issue out of a matter when voters tend to be in agreement. No candidate is going to say that he favors dirty air and polluted water, wants to see more dolphins killed, or hopes to build a Wal-Mart in the middle of Yosemite."[35] As a result, environmental debate tends to be driven by strongly emotive symbols that are universally agreed upon—a pattern which does little to inform voters, and even less to influence policy in ways that matter.

Communicating Risk

Despite the political limits imposed on those who seek to strengthen environmental laws and regulations, the accomplishments of the movement itself, at least, seem certain. There is comfort to be found in the success of the environmental agenda beyond most reasonable expectations. Disjointed laws and outdated technologies may persist, but a variety of statistical indicators relating to air, water, and land prove that our natural surroundings are cleaner and healthier than they were a generation ago because of considerable and combined efforts. But the bad news is that most Americans are unwilling to believe it.[36] Whether because of the news media's tendency toward pessimism or the reluctance of activists to concede success, a majority of those responding to one recent poll thought that the environmental movement had made only "minor" progress in pursuit of its goals. Sixteen percent insisted that environmental conditions had actually grown *worse* over the past thirty years.[37]

Popular views on subjects such as these are, according to Norman Levitt, "obstinately impermeable to scientific good sense," creating problems for experts whose job it is to communicate risk to the general public.[38] For example, since public opinion unsupported by facts fails to conform to the language of traditional policy analysis, which focuses on balancing quantifiable costs and benefits, scientists tend to view environmental values with an "uncomfortability bordering on disdain."[39] Often, when faced with disagreement, advocates attempt to persuade (or even coerce) people into altering their stubborn perceptions.[40] Under those conditions, it should come as no surprise that risk communication between experts and the lay public is plagued by distrust on both sides.

The answer to that dilemma, according to Supreme Court justice Stephen Breyer, author of *Breaking the Vicious Circle* (1993), is to be found in an insulated science, one protected from the ill-informed demands of citizens who exert pressure on lawmakers to allocate scarce funds in all the wrong directions.[41] Yet as proponents of civic environmentalism argue, "in many cases the decision to act on environmental issues cannot be solved by technocratic means" alone. If building public confidence and trust is as important as Breyer insists, perhaps science should not "displace democratic participation" but rather "situate it" and "provide it with a realistic context," using expertise to frame issues in ways average citizens can understand.[42]

Lessons from the Marketplace

While confusion over the science and politics of the environment continues, surprising evidence suggests that Americans are motivated in the marketplace by a growing environmental awareness and a willingness to vote at the cash register (if not always the ballot box) for improved environmental quality. Ironically, perhaps it is this "rise of a new consumer" that helps us best in understanding the conditions under which environmental preferences shape mass behavior and the opportunities under which those actions can be extended.[43]

For example, if green consumerism is driven by narrow self-interest, at least on a limited subset of issues like energy and resource conservation, perhaps the base motive of self-interest could be expanded into other areas with similar effect. If a concern for energy conservation or airborne pollutants demands that drivers economize on gasoline usage, Charles Schultze argues that

Warnings about the energy crisis, and "don't be fuelish" slogans are no match for higher prices at gas pumps. In most cases the prerequisite for social gains is the identification not of villains and heroes but of the defects in the incentive system that drive ordinary decent citizens into doing things contrary to the common good.[44]

In practice, that logic might extend not only to gasoline taxes but also to bottle bills and unit pricing for residential waste disposal. Communities such as Seattle, Washington, and Loveland, Colorado, have found that rather than charging a flat rate through property taxes, homeowners can be charged for the volume of trash they dispose, similar to fees for other public utilities. Once individual households see the link between what they throw away and what they pay, the hope is that solid waste will decrease, recycling will increase, and that consumers will buy products with less packaging in the first place.[45]

Second, if Americans prefer to buy green rather than vote green, perhaps mainstream environmental groups should become more involved in promoting the possibilities of consumer choice. That they do not seems to reflect the ideological divisions that the market so neatly avoids, at least as far as average consumers are concerned. Liberal environmental theory stresses government intervention as the preferred ethical solution to market failures on public goods.[46] It is not surprising, then, that Richard Ellis and Fred Thompson find that many elite-level environmental activists "do not trust markets" and that they often "spurn policies that rely on markets to solve problems."[47] Yet without an active environmental voice serving as watchdog-protector, consumers are left to rely on the environmental product information provided to them by corporations that exaggerate claims to improve profitability.

Third, if market choice is associated with a stronger sense of personal effectiveness, perhaps those feelings could be fostered better in the arena of politics. In *Green Culture* (1996), Carl Herndl and Stuart Brown suggest that many of the recruitment techniques used by environmental organizations belie the grassroots reputations those groups enjoy.[48] For instance, in response to the direct-mail solicitations that have grown popular in recent years, they say, one can either ignore the letter or opt to send money in support of some environmental cause. But the request itself generally offers individuals few options for true and valued participation—the kind of participation that allows consumers to feel good about their environmental efforts in the marketplace.

Finally, the persuasiveness of environmental labels in shaping (and sometimes altering) consumer preferences suggests that the way in which environmental issues translate into action may be contingent, ultimately, on the cues or symbols that are cognitively linked to it.[49] Given that voters often depend on political campaigns to simplify choice and inform electoral decisions, it is not surprising that environmental ballot measures falter when linked to broad symbols such as "big government," "bureaucracy" and "waste."[50] If environmentalists were to learn a lesson from Madison Avenue, perhaps it would be to more forcefully define the terms of debate for voters and candidates alike. Just as in the marketplace, political rhetoric can be key in shaping public willingness to pay for costly environmental reform.

Beyond Consensus

V. O. Key once wrote that the idea of consensus was a "handy crutch" for those who seek to describe patterns of public opinion. But it was, he thought, after much consideration, a term far less useful for that task than it might appear.[51] To take public opinion polls at face value, we would likely conclude that environmental attitudes approach a national consensus, one so broad in fact that inaction becomes socially and politically unimaginable. In reality, however, our collective desire to address environmental problems has far exceeded the ability of political leaders to do so.[52] Understanding the reasons why brings us full circle.

First, the politics of consensus serves to simplify complex environmental issues and conceal fundamental disagreement over how best to achieve those goals. Elevated to the status of a "valence" issue, dialogue tends to focus on shared values, while muting cleavages along ideological lines that make environmental policies divisive.[53] With genuine differences in opinion that compete over how best to address environmental problems, the fact that the environment has become a "political substitute for motherhood" invites little more than empty symbolism and "feel-good" rhetoric that ignores the conflicts and painful trade-offs created.[54] As Ted Gup writes in reflecting on the prolonged battle in the Pacific Northwest over the northern spotted owl, environmental and economic concerns are not always incompatible, but "the longer society lacks the political courage to act, the harder it is to find a solution."[55]

Second, environmental issues that appear to transcend politics displace decision-making to an elite level, where a different distribution of values and ideologies obtain.[56] Widespread public support surely lends credibility to the claim that environmentalists represent the public interest and it does provide an image of strength when lobbyists argue for new and stronger regulations. But in Key's sense of the term, broad agreement on environmental goals is likely to forge a "permissive consensus," where the absence of popular dissent and electoral reprisal allows political leaders to exercise a considerable degree of latitude in designing environmental policies free from a watchful public eye.[57]

Finally, and perhaps most important of all, the politics of consensus makes it possible for political actors to speak the language of environmentalism while serving their constituents in less authentic ways. Republican presidential candidate George W. Bush's speeches on the importance of the national parks during the 2000 campaign were "symbolically perfect" and "emotionally charged," according to one columnist, even though Bush's brand of environmentalism was, in his opinion, "rather like a Save the Planet sticker on a gas-guzzling SUV."[58]

In the end, the fact that politicians "greenwash" questionable environmental credentials, and corporations exaggerate environmental claims on product labels is troubling, not so much because it occurs, but because it succeeds. It is unsettling, not only because emotive symbols invoked through advertising and political campaigns have the power to persuade as well as to mislead, but also because, quite simply, "the Earth does not benefit from symbolic gestures."[59] Ultimately, the best advice this book has to offer is to resist the false appearance of consensus in favor of honest debate and tireless compromise. The result may fall short of revolution, but it might well be one that fosters public confidence in genuine environmental progress.

Appendix

A Note on Data Sources

Nearly all of the data used in this book come from public opinion polls administered to various state and national samples of American adults between 1973 and 2001. Because the quality of survey research (and the analysis on which it is based) is influenced by issues of sample size and selection, as well as question wording, format, and design, additional information on the data used in several chapters is detailed here.

Chapter 3

The variables used in figure 3.1 come from a wide variety of sources. Question wording is as follows:

Government spending on the environment "too little":

"We are faced with many problems in this country, none of which can be solved easily or inexpensively. I'm going to name some of these problems, and for each one I'd like you to tell me whether you think we're spending too much money on it, too little money, or about the right amount on ... improving and protecting the environment."

Source: NORC General Social Survey, 1973–1998.

Willing to sacrifice economic growth for environmental protection:

"Which of these two statements is closer to your opinion? We must sacrifice economic growth in order to preserve and protect the environment. Or, we must be prepared to sacrifice environmental quality for economic growth?"

Source: Cambridge Reports Research International data from 1976 to 1990 are drawn from Dunlap and Scarce, "The Polls," 664; data for 1991 to 1994 come directly from Cambridge Reports Research International *National Omnibus* data-files (September 1991, September 1992, September 1993, and September 1994).

Support environmental protection "regardless of cost":

"Do you agree or disagree with the following statement? Protecting the environment is so important that requirements and standards cannot be too high, and continuing environmental improvements must be made regardless of cost?"

Source: NYT/CBS data from 1981 to 1990 are drawn from Dunlap and Scarce, "The Polls," 664; data from 1992 to 1999 come from Wirthlin Worldwide datafiles.

Government regulation on the environment "too little":

"In general, do you think there is too much, too little, or about the right amount of government regulation and involvement in the area of environmental protection?"

Source: Cambridge Reports Research International data from 1982 to 1990 are drawn from Dunlap and Scarce, "The Polls," 664; 1991–1994 data come from Cambridge Reports Research International datafiles (September 1991, September 1992, September 1993, and September 1994); 1996 to 1999 data come directly from Wirthlin Worldwide datafiles.

Note: Wording used by Wirthlin Worldwide drops the phrase "In general" from the beginning of the question.

Environmental laws and regulations not gone far enough:

"There are also different opinions about how far we've gone with environmental protection laws and regulations. At the present time, do you think environmental protection laws and regulations have gone too far, or not far enough, or have struck about the right balance?"

Source: Roper data from 1973 to 1990 are drawn from Dunlap and Scarce, "The Polls," 664; 1992 data come directly from Roper datafiles.

Chapter 4

Variables used in tables 4.1, 4.2, and 4.4 are drawn from the 1994 NORC General Social Survey. The National Opinion Research Center (NORC) has administered its General Social Survey (GSS) either yearly or biennially since 1972. Datafiles are available through the Roper Center for Public Opinion Research at the University of Connecticut (⟨http://www.ropercenter.uconn.edu⟩) or the Inter-University Consortium for Political and Social Research (ICPSR) at the University of Michigan (⟨http://www.icpsr.umich.edu/GSS⟩).

Question wording and coded responses (in parentheses) are as follows. Original variable names appear in brackets.

Dependent Variables

General perceptions of environmental danger:

An additive scale ranging in value from 6 to 30 summed across the items below using this question, recoded as follows:

"In general, do you think that _____ is (5) Extremely dangerous for the environment; (4) Very dangerous; (3) Somewhat dangerous; (2) Not very dangerous; (1) Not dangerous at all for the environment?"

Personal perceptions of environmental danger:

An additive scale ranging in value from 6 to 30 summed across the items below using this question, recoded as follows:

"And do you think _____ is (5) Extremely dangerous for you and your family; (4) Very dangerous; (3) Somewhat dangerous; (2) Not very dangerous; (1) Not dangerous at all for you and your family?"

• Air pollution caused by cars [CARSGEN; CARSFAM]
• Nuclear power stations [NUKEGEN; NUKEFAM]
• Air pollution caused by industry [INDUSGEN; INDUSFAM]
• Pesticides and chemicals used in farming [CHEMGEN; CHEMFAM]
• Pollution of America's rivers, lakes and reservoirs [WATERGEN; WATERFAM]
• A rise in the world's temperature caused by the "greenhouse effect" [TEMPGEN; TEMFAM]

Support for increased environmental spending:

"We are faced with many problems in this country, none of which can be solved easily or inexpensively. I'm going to name some of these problems, and for each one I'd like you to tell me whether you think we're spending too much money on it, too little money, or about the right amount on.... Improving and protecting the environment." [NATENVIR]

Recoded as (1) Too much; (2) About the right amount; (3) Too little.

Willingness to pay higher taxes:

"And how willing would you be to pay much higher taxes in order to protect the environment?" [GRNTAXES]

Recoded as (1) Not at all willing; (2) Not very willing; (3) Neither willing nor unwilling; (4) Fairly willing; (5) Very willing.

Economic behavior:

An additive scale ranging in value from 3 to 12 summed across the following items:

• "How often do you make a special effort to sort glass or cans or plastic or papers and so on for recycling?"
Recoded as (1) Never; (2) Sometimes; (3) Often; (4) Always. [RECYCLE]
• "And how often do you make a special effort to buy fruits and vegetables grown without pesticides or chemicals?"
Recoded as (1) Never; (2) Sometimes; (3) Often; (4) Always. [CHEMFREE]
• "And how often do you cut back on driving a car for environmental reasons?"
Recoded as (1) Never; (2) Sometimes; (3) Often; (4) Always. [DRIVLESS]

Political activism:

An additive scale ranging in value from 0 to 4 summed across the following items:

• "Are you a member of any group whose main aim is to preserve or protect the environment?"
Coded as (0) No; (1) Yes. [GRNGROUP]
• "In the last five years, have you signed a petition about an environmental issue?"
Coded as (0) No; (1) Yes. [GRNSIGN]
• "[In the last five years, have you] given money to an environmental group?"
Coded as (0) No; (1) Yes. [GRNMONEY]
• "[In the last five years, have you] taken part in a protest or demonstration about an environmental issue?"
Coded as (0) No; (1) Yes. [GRNDEMO]

Independent Variables

Party identification:
"Generally speaking, do you usually consider yourself a Republican, Democrat, Independent, or what?"
Recoded as (0) Strong Republican; (1) Not very strong Republican; (2) Independent, close to Republican; (3) Independent; (4) Independent, close to Democrat; (5) Not very strong Democrat; (6) Strong Democrat. [PARTYID]

Political ideology:
"We hear a lot of talk these days about liberals and conservatives. I'm going to show you a seven-point scale on which the political views that people might hold are arranged from extremely liberal—point 1—to extremely conservative—point 7. Were would you place yourself of this scale?"
Recoded as (1) Extremely conservative; (2) Conservative; (3) Slightly conservative; (4) Moderate, middle-of-the-road; (5) Slightly liberal; (6) Liberal; (7) Extremely liberal. [POLVIEWS]

Age:
Respondent's age coded in years, then converted into birth cohorts. [AGE]

Education:
Respondent's level of education in total years. [EDUC]

Income:
Respondent's family income coded into 21 categories. [INCOME91]

Race:
Respondent's race, where (1) Black; (2) Other; (3) White. [RACE]

Gender:
Respondent's gender, where (0) Female; (1) Male. [SEX]

Chapter 6

The data used in chapters 4 and 6 are drawn from the 1996 National Election Study, Pre- and Post-Election Surveys (ICPSR 6896), which are available through the Center for Political Studies of the Institute for Social Research at the University of Michigan (⟨http://www.umich.edu/~nes⟩). A random sample of 1,523 American adults was interviewed nationwide from September 3–November 4, 1996, and again from November 6–December 31, 1996.

Question wording and coded responses (in parentheses) are as follows. Original variable names appear in brackets.

Dependent Variables

Feeling thermometers:

"I'd like to get your feelings toward some of our political leaders and other people who are in the news these days. I'll read the name of a person and I'd like you to rate that person using something we call the feeling thermometer. Ratings between 50 degrees and 100 degrees mean that you feel favorable and warm toward the person. Ratings between 0 and 50 degrees mean that you don't feel favorable toward the person and that you don't care too much for that person. You would rate the person at the 50 degree mark if you don't feel particularly warm or cold toward the person. If we come to a person whose name you don't recognize, you don't need to rate the person. Just tell me and we'll move on to the next one."

• *Environmentalists Feeling Thermometer:* "How would you rate environmentalists?" [v961041]
• *Clinton Feeling Thermometer:* "How would you rate Bill Clinton?" [v960272]
• *Dole Feeling Thermometer:* "How would you rate Bob Dole?" [v960273]

Independent Variables

Partisan identification:

Summaries of respondent's partisan identification based on the following series of questions:

"Generally speaking, do you usually think of yourself as a Republican, a Democrat, an Independent, or what?"

[If considers self Republican/Democrat:] "Would you call yourself a strong [Republican/Democrat] or a not very strong [Republican/Democrat]?"

[If considers self independent/no preference/other:] "Do you think of yourself as closer to the Republican Party or to the Democratic Party?"

Final scale is coded as (0) Strong Democrat; (1) Weak Democrat; (2) Independent-Democrat; (3) Independent; (4) Independent-Republican; (5) Weak Republican; (6) Strong Republican. [v960420]

Political ideology:

"We hear a lot of talk these days about liberals and conservatives. Here is a seven-point scale on which the political views that people might hold are arranged from extremely liberal to extremely conservative. Where would you place yourself on this scale, or haven't you thought much about this?"

Coded as (1) Extremely liberal; (2) Liberal; (3) Slightly liberal; (4) Moderate; middle of the road; (5) Slightly conservative; (6) Conservative; (7) Extremely conservative. [v960365]

Environment/economy:

"Some people think it is important to protect the environment even if it costs some jobs or otherwise reduces our standard of living. (Suppose these people are at one end of the scale, at point number 1.) Other people think that protecting the environment is not as important as maintaining jobs and our standard of living. (Suppose these people are at the other end of the scale, at point number 7.) And, of course, some other people have opinions somewhere in between, at points 2, 3, 4, 5, or 6. Where would you place yourself on this scale, or haven't you thought much about this?" [v960523]

Importance of environment/economy:

"How important is this issue to you?" [v960525]

Coded as (1) Not important at all; (2) Not too important; (3) Somewhat important; (4) Very important; (5) Extremely important.

Government health insurance:

"There is much concern about the rapid rise in medical and hospital costs. Some people feel there should be a government insurance plan that would cover all medical and hospital expenses for everyone. (Suppose these people are at one end of a scale, at point 1). Others feel that all medical expenses should be paid by individuals and through private insurance plans like Blue Cross or some other company paid plans. (Suppose these people are at the other end, at point 7). And, of course, some other people have opinions somewhere in between at points 2, 3, 4, 5, or 6. Where would you place yourself on this scale, or haven't you thought much about this?" [v960479]

Guaranteed job/standard of living:

"Some people feel the government in Washington should see to it that every person has a job and a good standard of living (Suppose these people are at one end of a scale, at point 1.) Others think the government should just let each person get ahead on their own. (Suppose these people are at the other end, at point 7.) And, of course, some other people have opinions somewhere in between, at points 2, 3, 4, 5, or 6. Where would you place yourself on this scale, or haven't you thought much about this?" [v960483]

Services/spending:

"Some people think the government should provide fewer services, even in areas such as health and education, in order to reduce spending. (Suppose these people are at one end of a scale, at point 1.) Other people feel it is important for the government to provide many more services, even if it means an increase in spending. (Suppose these people are at the other end, at point 7.) And, of course, some other people have opinions somewhere in between, at points 2, 3, 4, 5, or 6. Where would you place yourself on this scale, or haven't you thought much about this?"

Note: In order to maintain ideological consistency with the other scales used here, this item was recoded in reverse order. [v960450]

Aid to blacks:

"Some people feel that the government in Washington should make every effort to improve the social and economic position of blacks. (Suppose these people are at one end of a scale, at point 1.) Others feel that the government should not make any special effort to help blacks because they should help themselves. (Suppose these people are at the other end, at point 7.) And, of course, some other people have opinions somewhere in between, at points 2, 3, 4, 5, or 6. Where would you place yourself on this scale, or haven't you thought much about this?" [v960487]

Reduce crime:

"Some people say that the best way to reduce crime is to address the social problems that cause crime, like bad schools, poverty, and joblessness. (Suppose these people are at one end of a scale, at point 1.) Other people say the best way to reduce crime is to make sure that criminals are caught, convicted, and punished. (Suppose these people are at the other end, at point 7.) And, of course, some other people have opinions somewhere in between, at points 2, 3, 4, 5, or 6. Where would you place yourself on this scale, or haven't you thought much about this?" [v960519]

Women's rights:

"Recently there has been a lot of talk about women's rights. Some people feel that women should have an equal role with men in running business, industry, and government. (Suppose these people are at one end of a scale, at point 1.) Others feel that a woman's place is in the home. (Suppose these people are at the other end, at point 7.) And, of course, some other people have opinions somewhere in between, at points 2, 3, 4, 5, or 6. Where would you place yourself on this scale, or haven't you thought much about this?" [v960543]

Defense spending:

"Some people believe that we should spend much less money for defense. (Suppose these people are at one end of a scale, at point 1.) Others feel that defense spending should be greatly increased. (Suppose these people are at the other end, at point 7.) And, of course, some other people have opinions somewhere in between, at points 2, 3, 4, 5, or 6. Where would you place yourself on this scale, or haven't you thought much about this?" [v960463]

Abortion rights:

"There has been some discussion about abortion during recent years. Which one of the opinions on this page best agrees with your view? You can just tell me the number of the opinion you choose: (1) By law, abortion should never be permitted; (2) The law should permit abortion only in case of rape, incest, or when the woman's life is in danger; (3) The law should permit abortion for reasons other than rape, incest, or danger to the woman's life, but only after the need for the abortion has been clearly established; (4) By law, a woman should always be able to obtain an abortion as a matter of personal choice." [v960503]

State of the nation's economy:

"Now thinking about the economy in the country as a whole, would you say that over the past year the nation's economy has (1) Gotten better; (3) Stayed about the same; or (5) Gotten worse?" [v960385]

Age:

"What is the month, day and year of your birth?" [v960605]

The month and year of the respondent's birth was subtracted from the month and year of the interview to calculate age in years.

Education:

Summary of respondent's level of education [v960610], where (1) 8 grades or less and no diploma or equivalency; (2) 9–11 grades, no further schooling; (3) High school diploma or equivalency test; (4) More than 12 years of schooling, no higher degree; (5) Junior or community college level degree; (6) Bachelor's level degree; (7) Advanced degree.

Income:

"Please look at page 21 of the booklet and tell me the letter of the income group that includes the income of all members of your family living here in 1995 before taxes. This figure should include salaries, wages, pensions, dividends, interest, and all other income." [v960701]

Coded into 24 categories, ranging in value from a low of "none or less than $2,999" to a high of "$105,000 and over."

Gender:

Respondent's gender [v960066] is recoded as (0) Female; (1) Male.

Race:

Respondent's race [v960067] is recoded as (0) White; (1) Black/American Indian/ Alaskan native/Asian or Pacific Islander/Other.

Chapter 7

The data used in figure 7.4 and table 7.3 are drawn from a series of California polls administered in 1986 by the Field Institute. Those polls are available through the University of California and the Roper Center for Public Opinion Research at the University of Connecticut (⟨http://www.ropercenter.uconn.edu⟩). Adult residents of California were selected by random-digit dialing and then interviewed by telephone. The first poll (conducted July 24–August 4, 1986) had a sample size of 1,028 respondents; the second poll (September 4–October 2) had a sample size of 1,028; and the third poll (October 29–30) had a sample size of 701. For more information, contact the Field Institute at: 550 Kearney Street, Suite 900, San Francisco, CA 94108-2527, or at ⟨http://www.field.com/fieldpoll⟩.

Question wording and coded responses (in parentheses) are as follows.

Dependent Variables

Vote intention on Proposition 65:

"Proposition 65 would prohibit the discharge of toxic chemicals into drinking water and require warnings of toxic chemicals exposure. If you were voting today on Proposition 65, would you vote yes or no?"

Answers are coded as (1) Vote no; (2) Undecided; (3) Vote yes.

Independent Variables

Age:

"May I ask your age please?"

Coded in years.

Education:

"What is the highest grade or year of school that you have finished and gotten credit for?"

Answers are coded as (1) 8th grade or less; (2) Some high school; (3) Graduated from high school; (4) Trade/vocational school; (5) 1–2 years of college; (6) 3–4 years of college; (7) Graduated from college; (8) 5–6 years of college; (9) Master's degree; (10) Post-master's.

Income:

"Now, we don't want your exact income, but just roughly could you tell me if your annual household income before taxes is ..."

Answers are coded as (1) Under $10,000; (2) $10,000–$19,999; (3) $20,000–$29,999; (4) $30,000–39,999; (5) $40,000–$49,999; (6) $50,000–$59,999; (7) $60,000–$69,999; (8) $70,000 or more.

Partisan identification:

"Generally speaking, in politics do you consider yourself a conservative, liberal, middle-of-the-road or don't you think of yourself in these terms?"

[If liberal/conservative:] "Do you consider yourself a strong or not very strong liberal/conservative?"

[If middle-of-the-road:] "Do you think of yourself as closer to conservatives or closer to liberals?"

Answers to this series of questions are coded as (1) Strong conservative; (2) Not a strong conservative; (3) Moderate, closer to conservative; (4) Moderate; (5) Moderate, closer to liberal; (6) Not a strong liberal; (7) Strong liberal.

Political ideology:

"Generally speaking, do you usually consider yourself as a Republican, a Democrat, and Independent, or what?"

[If Democrat/Republican:] "Would you call yourself a strong or not very strong Republican/Democrat?" [If Independent, no preference, other, or don't know:] "Do you think of yourself as closer to the Republican or the Democratic party?"

Answers to this series of questions are coded as (1) Strong Republican; (2) Not very strong Republican; (3) Independent, closer to Republican; (4) Independent; (5) Independent, closer to Democrat; (6) Not very strong Democrat; (7) Strong Democrat.

Race:

"For classification purposes, we'd like to know what your racial background is. Are you white, black, Asian, or are you a member of some other race?"

Respondent's race is coded here as (0) White; (1) Other.

Gender:

Respondent's gender, coded as (0) Female; (1) Male.

Prior knowledge of Prop. 65:

"Have you seen or heard anything about an initiative, Proposition 65, that will be on the November statewide election ballot having to do with toxic substances and exposure restrictions?"

Answers are coded as (0) Have not heard; (1) Have heard.

Data for figure 7.4, and tables 7.4 and 7.5 come from a 1992 poll conducted by Marttila & Kiley, Inc. Originally designed for the Massachusetts Public Interest Research Group (MassPIRG), the environmental sponsors of Question 3, the questionnaire (identified internally as #MK-92110) was administered by telephone February 11–13, 1992. A sample of 402 likely Massachusetts voters was generated using a random-digit dialing. The sample was stratified according to county, and gender quotas were observed. The raw datafile was provided to the author courtesy of Marttila Communications Group, Inc. (1 Beacon Street, Boston, MA 02108), with the express approval of MassPIRG (29 Temple Place, Boston, MA 02111).

Question wording and coded responses (in parentheses) for the variables used in this book are as follows. The original number of each item on the questionnaire is noted in brackets.

Dependent Variables

Vote intention #1:

"Please tell me whether your initial reaction would be to vote in favor of that proposal or vote against the proposal: A proposal that would require nearly all packaging used in the state to be recyclable or made of recycled materials." [Q11]

Coded as (1) No; (4) Yes.

Vote intention #2:

"Let me describe this proposal in a little more detail. Packaging accounts for roughly one-third of the total volume of trash disposal in Massachusetts each year. In order to sharply reduce the amount of trash, this proposal would require that by July 1, 1996, packaging will have to meet one of five standards, by being smaller in size, reusable, recycled, or made of recycled or recyclable materials. Manufacturers and businesses can use any one of the five standards to meet the new packaging requirements. If the election on this proposal were being held tomorrow, would you be inclined to vote yes or no on this proposal?"

[If not sure] "I know you could change your mind, but which way are you leaning based on this information?" [Q18]

Coded as (1) No; (2) Lean no; (3) Lean yes; (4) Yes.

Vote intention #3:

[After asking respondents to react to a lengthy list of positive and negative reasons for supporting/rejecting and proposal]. "Now that you have heard some of the practical concerns and reservations about the recycling initiative, I want to see how you feel now: If the election were held tomorrow, would you probably vote yes on this proposal, are you leaning toward voting yes, are you leaning toward voting no, or would you probably vote no on this proposal?" [Q36]

Coded as (1) Probably vote no; (2) Lean no; (3) Lean yes; (4) Yes.

Change in vote intention:

This variable measures change in vote intention on Question 3 after considering both positive and negative aspects of the proposal. By subtracting a respondent's answer to the second survey condition from their answer to the first, a scale was created that ranges in value from −3 to +3, where 0 indicates no change in voting intention between the second and third question. The presence of a negative sign means that a respondent's vote intention became less favorable, while a positive sign indicates that their vote intention grew more favorable.

Independent Variables

Age:

"In which category does your age fall?" [Q201]

Answers are coded responses as (1) Under 25; (2) 25 to 29; (3) 30 to 34; (4) 35 to 39; (5) 40 to 44; (6) 45–49; (7) 50 to 54; (8) 55 to 59; (9) 60 to 64; (10) 65 and over.

Education:

"What was the last grade of school you completed?" [Q202]

Answers are coded as (1) Grade school or less; (2) Some high school; (3) High school graduate; (4) Vocational/technical; (5) Some college/2-year college; (6) Four-year college graduate; (7) Post-graduate work.

Income:

"For tabulation purposes only, please tell me which of the following income categories includes your total family income in 1991 before taxes—just stop me when I read the correct category." [Q206]

Answers are coded as (1) Less than $20,000; (2) $20,000–$29,999; (3) $30,000–$39,999; (4) $40,000–$49,999; (5) $50,000–$74,999; (6) $75,000 or over.

Gender:

By observation, coded as (0) Female; (1) Male. [Q206]

Partisan identification:

"Regardless of which party you like better these days, are you currently registered to vote in Massachusetts as a Democrat, Republican, or an Independent?" [Q203]

Answers are coded as (1) Republican; (2) Independent; (3) Democrat.

Political ideology:

"When it comes to most political issues, do you think of yourself as a liberal, a conservative, or a moderate? (If moderate—Do you think of yourself as closer to being liberal or being conservative?)" [Q204]

Answers are coded (1) Conservative; (2) Moderate-conservative; (3) Moderate; (4) Moderate-liberal; (5) Liberal.

Household recycles:

"Other than returning bottles and cans for deposit, do you or does your household participate in a recycling program in your community, do you voluntarily recycle certain items even though it is not part of a community program, or isn't your household involved in recycling yet?" [Q15]

Responses are coded (1) Participate in community recycling program *or* voluntarily recycle without program; (0) Not involved in recycling.

Additive scale of positive evaluations:

Respondents were asked to react to a battery of nine statements, saying how important each reason was for voting in favor of the proposal: (1) Not very important; (2) Somewhat important; (3) Very important; or (4) Extremely important. Answers to each of these questions were summed into a scale ranging in value from 9 to 36. Reasons included, among others: "Jump-starting the economy by creating new recycling jobs"; "reducing our reliance on landfills"; and "saving money in trash disposal costs." [Q19–Q27]

Additive scale of negative evaluations:

Respondents were asked to react to a battery of eight statements, saying to what extent each reason gave them reservations about the proposal: (1) No reservations; (2) Minor reservations; (3) Fairly strong reservations; or (4) Very strong reservations. Answers to each of these questions were summed into a scale ranging in value from 8 to 32. Reservations included, among others: "Creating a whole new government bureaucracy to enforce new, complicated packaging standards"; "fewer choices for consumers in the supermarket"; "banning plastic packaging used to keep fresh foods fresh and sanitized"; and "job losses in plastics and packaging industries." [Q28–Q35]

Chapter 8

Data from the Omni Study cited in chapter 8 were provided to the author courtesy of Cambridge Reports Research International, 955 Massachusetts Avenue, Cambridge, MA 02139. The survey was administered July 15–27, 1993, to a national sample of 1,250 American adults.

Question wording and coded responses (in parentheses) for the variables used in figure 8.1 and table 8.2 are as follows.

Dependent Variables

Frequency of buying green products:

"Now I'm going to read a list of things people have done because of their concern about the environment. Please use a scale of '1' to '7,' where '1' means 'never have done that thing' and '7' means 'very frequently do that thing' to tell me how often you do each thing I read."

Individual survey items were coded from 1 to 7. An additive scale was created by summing responses across the following questions:

1. Recycle used cans, bottles or paper [Q21];
2. Vote for a political candidate primarily because the candidate took strong environmental positions [Q22];
3. Cut back on driving or use public transportation more often [Q23];
4. Buy products made of recycled material whenever possible [Q24];
5. Avoid purchasing certain kinds of products because the packaging is excessive or environmentally harmful [Q25];
6. Avoid purchasing certain kinds of fresh food because of the chemicals used in food production [Q26];
7. Avoid purchasing products made by a company that pollutes the environment [Q27];
8. Buy products in packages that can be refilled [Q28];
9. Buy a product because the label or advertising said it was environmentally safe or biodegradable [Q29];
10. Avoid restaurants using plastic foam containers [Q30];
11. Avoid buying products in aerosol containers [Q31].

Willingness to pay more for environmental products:

"Now I am going to read you a list of products. Please tell me how much more you would be willing to pay for each product if it were environmentally friendly. If you wouldn't be willing to pay anything more for the product if it were environmentally friendly, please just say so."

Individual survey items are coded as (0) Wouldn't pay anything more; (1) 5% more; (2) 10% more; (3) 20% more; (4) at least 30% more. The additive scale used as a dependent variable in table 8.2 was created by summing responses across the following questions:

1. Automobile [Q8];
2. Detergents [Q9];
3. Paper products made out of recycled paper [Q10];
4. Gasoline [Q11];
5. Plastic packaging or containers made of recycled plastic materials [Q12];
6. Plastic packages or containers made with less plastic [Q13];
7. Garden products such as insecticides and fertilizers [Q14];
8. Household products, such as kitchen and bathroom cleaners [Q15].

Independent Variables

Environmental self-identification:

"Using a scale of '1' to '7,' where '1' means 'do not identify with at all' and '7' means 'strongly identify with,' please tell me how much you identify yourself with the label 'environmentalist'." [Q48]

Age:

Respondent's age [Q53] as measured in categories, where (1) Ages 18 to 25; (2) Ages 26 to 35; (3) Ages 36 to 45; (4) Ages 46 to 55; (5) Ages 56 to 65; and (6) Over age 65.

Education:

Respondent's education [Q52] as measured in categories, where (1) Some grade school; (2) Some high school; (3) Graduated high school; (4) Technical/vocational school; (5) Some college; (6) Graduated college; and (7) Graduate/professional school.

Income:

"Would you please tell me in which of the categories I read is your total household income—of everyone living in the house?" [Q54].

Responses are coded as (1) $0 to $7,999; (2) $8,000 to $11,999; (3) $12,000 to $14,999; (4) $15,000 to $19,999; (5) $20,000 to $24,999; (6) $25,000 to $34,999; (7) $35,000 to $49,000; (8) $50,000 to $74,999; (9) $75,000 to $99,999; and (10) $100,000 and over.

Political ideology:

"Would you describe yourself as more of a liberal or more of a conservative?" [Q55].

Responses are coded as (1) Conservative; (2) Moderate; (3) Liberal.

Race:

Respondent's race [Q51] is coded dichotomously, where (0) white and (1) black/other.

Gender:

Respondent's gender [Q50] is coded as (0) female and (1) male.

Notes

Introduction

1. Key, *Public Opinion and American Democracy*, 550.

2. Rubin, *The Green Crusade*.

3. The most famous example of children's environmental literature remains Dr. Seuss's cautionary tale of the *The Lorax* (1971), in which the title character "speaks for the trees" when a rapacious industrialist sets out to destroy a forest of Truffula trees, leading to widespread environmental devastation. For more, see Sinclair, *E for Environment*.

4. The television show *Murphy Brown* was notable for showing its major characters recycling bottles and cans in the workplace. *The Simpsons* is also well known for its environmental themes, including the dangers of nuclear power in an episode titled "Two Cars in Every Garage, Three Eyes on Every Fish." For a discussion of more subtle environmental cues, see Bernstein, "Planet Hollywood," 4; Drum, "Product Placement Matures into Placement of Nonprofit Causes," S27; and Leroux, "Subliminal Messages," 14.

5. Two major theatrical releases with similar themes, *Erin Brockovich* (2000) and *A Civil Action* (1998), draw attention to the health risks associated with environmental pollution—the former, to illegal corporate dumping of toxic waste and the resulting pollution of groundwater; the latter, to the effects of a chemical spill near Woburn, Massachusetts.

6. Fenly, "New Books Tell Children about Environment and Their Future," 5.

7. Key, *Public Opinion*, 27.

8. Saad and Dunlap, "Americans Are Environmentally Friendly, but Issue Not Seen as Urgent Problem," 12–18.

9. Dowie, *Losing Ground*.

10. Rothman, *The Greening of a Nation?*, 210.

11. In fairness to both sides, however, Rubin's criticism should be placed to some extent in an ideological context. His view of early environmental writers

may well reflect his conservative approach to environmental policy. See, Rubin, *The Green Crusade*, 74.

12. In a Wirthlin Worldwide poll administered in August 1997, 41 percent of respondents agreed that environmental groups overstate threats in order to win public attention.

13. Dowie, *Losing Ground*; Easterbrook, *A Moment on the Earth*; Lomborg, *The Skeptical Environmentalist*.

14. David Brower, as quoted in Martin, "Earth Day Report Card—We Still Care, Sort of," A1.

15. In the public relations battle waged between optimism and pessimism on the future of the environment, pessimism seems to be winning. In the Gallup Organization's April 2000 public opinion poll, 64 percent of those surveyed believed that "only some progress" had been made in dealing with environmental problems over the past few decades. See Saad and Dunlap, "Americans Are Environmentally Friendly"; see also Meadows, "Looking Back at 30 Earth Days," B7.

16. Fried, "U.S. Environmental Interest Groups and the Promotion of Environmental Values," 1–2.

17. One of the must public controversies surrounding the celebration of Earth Day 2000 involved the appointment of actor Leonardo DiCaprio as honorary national chairman. The decision drew attention to a younger audience but also led to distractions when Vice President Al Gore's speech was punctuated by the screams of teenage girls eager to catch a glimpse of the *Titanic* star behind the stage. The actor later interviewed President Clinton during a live television broadcast, much to the consternation of seasoned news reporters. The special, *Planet Earth 2000*, ranked fourth in the Nielson ratings during its time slot. See Soskis, "Green with Envy," 13; Jensen, "DiCaprio and 'Planet Earth' Rank Fourth in Time Slot," F2.

18. For background, see Russell, "Corporations Going for the Green," D1; and Begley, Hager, and Wright, "The Selling of Earth Day," 60. For a harsher opinion piece, see Carothers, "Unwelcome Saviors," 62.

19. Rothman, *The Greening of a Nation?*, 210.

20. Mitchell, "Public Opinion and Environmental Politics in the 1970s and 1980s," in Vig and Kraft, eds., *Environmental Policy in the 1980s*.

21. Stanfield, "Reagan's Environmental Record Not the Issue It Was Predicted to Be," 1872.

22. Dunlap, "Polls, Pollution, and Politics Revisited," 34.

23. Vig, "Presidential Leadership and the Environment," 71.

24. Harris and Nakashima, "Gore's Greenness Fades," A1.

25. "How Green Is Al Gore?," 30.

26. See, for example, Mastio, "The GOP's Enviro-Rut," 19.

27. Knickerbocker, "Americans Go 'Lite Green,'" 1.

28. Newt Gingrich, quoted in Connelly, "The Politics of the Environment," H1. See also "G.O.P. Hears Nature's Call," and Schroeder, "Eco-Pragmatism." While Republicans have backed away from several obvious proposals to scale back environmental regulations on business, party members still attach anti-green riders to large-scale spending bills in an attempt to change policy with a minimum of debate (see, for example, Ourlian, "Anti-Green Measures Are Riders in the Storm."

29. Breyer, *Breaking the Vicious Circle*, 3.

30. Breyer, *Breaking the Vicious Circle*, 35.

31. Slovic, "Perception of Risk."

32. Rauch, "There's Smoke in the Air, But it Isn't All Pollution," B1.

33. See, for example, the editorial "Green Choices, Hard Choices."

34. A comprehensive guide to green products can be found online at ⟨http://www.buygreen.com⟩.

35. Schwartz and Miller, "The Earth's Best Friends"; Coddington, *Environmental Marketing*; Stisser, "A Deeper Shade of Green."

36. Wasik, "Market Is Confusing, But Patience Will Pay Off," 16; Morris, Hastak, and Mazis, "Consumer Comprehension of Environmental Advertising and Labeling Claims."

37. Ottman, "Environmentalism Will Be the Trend of the '90s," 13.

38. Hume, "Consumer Doubletalk Makes Companies Wary," GR4; Mantese, "Study Finds Green Confusion," 1; Sims, "Positive Attitudes Won't Make Cash Register Ring," 4.

39. In *Green Culture* Herndl and Brown suggest that many of the recruitment techniques used by environmental organizations belie the grassroots reputations those groups enjoy. For instance, in response to the direct-mail solicitations that have grown popular in recent years, they say, one can either ignore the letter or opt to send money in support of some environmental cause. But the request itself generally offers individuals few options for true and valued participation. Friedman, too, finds that environmental groups that pursue product boycotts are often more concerned about generating media attention than about acting in the marketplace. See also Friedman, "On Promoting a Sustainable Future through Consumer Activism."

40. Ridley and Low, "Can Selfishness Save the Environment?," 70.

41. Shabecoff, *Earth Rising*.

42. Bosso, "After the Movement," in Vig and Kraft, eds., *Environmental Policy in the 1990s*, 33; see also Dunlap, "Trends in Public Opinion toward Environmental Issues."

43. Breyer, *Breaking the Vicious Circle*, 39.

44. For a particularly thorough look at the subject, see also Morrison and Dunlap, "Environmentalism and Elitism"; and Van Liere and Dunlap, "The Social Bases of Environmental Concern."

45. Tucker, *Progress and Privilege*.

46. Sagoff, "The Great Environmental Awakening."

47. Gelernter, "The Immorality of Environmentalism," 14.

48. Switzer, *Green Backlash*.

49. Kempton, Boster, and Hartley, *Environmental Values in American Culture*, 211.

50. Ellis and Thompson, "Culture and the Environment in the Pacific Northwest."

51. Lake, "The Environmental Mandate," 232.

52. Schneider, "Everybody's An Environmentalist Now," 1062.

53. Dunlap, "Polls, Pollution, and Politics Revisited."

54. Shabecoff, *Earth Rising*, 113.

55. Key, *Public Opinion*, 8.

56. Schneider, "Everybody's An Environmentalist Now," 1062.

57. Key, *Public Opinion*, 232. See also Schumann and Presser, *Questions and Answers in Attitude Surveys*; Bryce, "The Nature of Public Opinion," and Lowell, "Public Opinion," both in Janowitz and Hirsch, eds., *Reader in Public Opinion and Mass Communication*.

58. Maslow, *Motivation and Personality*; Inglehart, "Post-Materialism in an Environment of Insecurity"; Dunlap, "Public Opinion and Environmental Policy," in Lester, ed., *Environmental Politics and Policy*; Downs, "Up and Down with Ecology."

59. See, for example, Kempton, Boster, and Hartley, *Environmental Values in American Culture*.

60. Key, *Public Opinion*.

61. deHaven-Smith, "Environmental Belief Systems," "Toward a Communicative Theory of Environmental Opinion," and *Environmental Concern in Florida and the Nation*.

62. The term *constraint* was first introduced in Converse, "The Nature of Belief Systems in Mass Publics," in Apter, ed., *Ideology and Discontent*.

63. Key, *Public Opinion*, 28.

64. Sagoff, "The Great Environmental Awakening."

Chapter 1

1. Gerstenzang, "Survey Bolsters Global Warming Fight," A4. Fifty-five percent of respondents in the Pew Research Center poll—the "majority" referred to in the text—supported a U.S. joint effort with other countries "in setting standards to improve the global environment." An additional 41 percent believed that the

United States should "set its own environmental standards independently." See The Pew Research Center for the People & the Press, *November 1997 News Interest Index*, Q23.

2. For example, in addition to stories in several major American newspapers, the results of the Pew Research Center poll were discussed on National Public Radio (NPR) and picked up by the Associated Press. See Soto, "Many Favor an Increase in Gas Prices."

3. Klein, "Environment Challenges Public Mood," 4. In "Culture and the Environment in the Pacific Northwest," Ellis and Thompson, too, reinforce the obvious—that activists and the general public "want cleaner air and water, wilderness and species preservation, habitat protection, and a healthier, safer earth" because these are so obviously desirable (p. 892). It is a natural tendency that creates problems for pollsters.

4. Martin, "Environment Is a Big Concern for Californians, Poll Shows," A3.

5. Alpert, "Voters Support Rules on Pollution," A1.

6. Dawson, "Poll Shows Texans Wanting to Recycle," 12.

7. Eddy, "Wilderness Expansion Backed," B6.

8. Richard Darman, quoted in Sagoff, "The Great Environmental Awakening," 39, emphasis added.

9. Schneider, "Everybody's an Environmentalist Now," 1062.

10. See, for example, Bowman, "The Rise of Environmentalism."

11. Kempton, Boster, and Hartley, *Environmental Values in American Culture*, 211.

12. Kempton, Boster, and Hartley, *Environmental Values*, 202–203.

13. Kempton, Boster, and Hartley, *Environmental Values*, Q31.

14. Kempton, Boster, and Hartley, *Environmental Values*, 114.

15. Saad and Dunlap, "Americans Are Environmentally Friendly."

16. Kempton, Boster, and Hartley, *Environmental Values*, 211. See also La Trobe and Acott, "A Modified NEP/DSP Environmental Attitudes Scale."

17. Kempton, Boster, and Hartley, *Environmental Values*, 209.

18. Ellis and Thompson, "Culture and the Environment in the Pacific Northwest," 892.

19. Kempton, Boster, and Hartley, *Environmental Values*, 203.

20. A number of measures on the lengthy Kempton questionnaire did probe respondents to consider policy options or economic costs and trade-offs, but those statements were far more likely to yield disagreement. For instance, when confronted with these statements: "My first duty is to feed my family. The environment and anything else has to come after that"—70 percent of sawmill workers agreed and just 13 percent of Earth First! members agreed. See Taylor, "Environmental Values in American Culture."

21. More information about the Consultative Group on Biological Diversity can be found at its Web site at ⟨http://www.cgbd.org/⟩.

22. Belden and Russonello, *The Ecology*, Q6, Q23, combined responses, "very important," "somewhat important."

23. For example, in an April 2000 Gallup poll, the seriousness of environmental problems ranked fifth among seven issues, behind drug use, crime and violence, health care, and homelessness.

24. See, for example, Traugott and Lavrakas, *The Voter's Guide to Election Polls*, especially chapter 7.

25. See Williams, "Window to the Wild."

26. Schuman and Presser, *Questions and Answers in Attitude Surveys*; also Sudman and Bradburn, *Asking Questions*.

27. Louis Harris and Associates, Inc., *Public and Leadership Attitudes to the Environment in Four Continents*.

28. "Worldwide Concern about the Environment," 14–15.

29. World Wildlife Fund, *Voters Believe Global Warming is a Reality*.

30. World Wildlife Fund, "State of the Climate," Q6, which read:

When cars, electric utilities and some other industries burn oil, coal and gasoline they release carbon dioxide and other gases into the atmosphere. Some scientists say these gases trap heat in the atmosphere, causing the earth to warm. They say this makes the world hotter and increases the chances of all kinds of extreme weather. This effect is called *global warming*. Generally speaking, how serious of a threat do you think global warming is today—very serious, somewhat serious, not too serious, or not serious at all? Or don't you have an opinion on this?

31. World Wildlife Fund, "State of the Climate."

32. Most respondents chose to answer these questions rather than rely (as they were reminded they could) on a "don't know" response. This is not surprising, however, when placed in broader context. Research has long shown that people can feel pressured to give responses in a survey context, even on fictitious issues (see, for example, Bishop, Tuchfarber, and Oldendick, "Opinions on Fictitious Issues").

33. Immerwahr, *Waiting for a Signal*. Apparently the public is not alone in its state of confusion. Politicians and newspaper columnists are also frequently misinformed. See Harris, "Experts in Everything and Nothing."

34. The Pew Research Center for the People & the Press, *November 1997 News Interest Index*, Q7.

35. Combined responses to the variable GRNTEST3 for "definitely true" and "probably true" (NORC General Social Survey 1994).

36. The Pew Research Center for the People & the Press, *November 1997 News Interest Index*, Q4, a–k.

37. Traugott and Lavrakas, *The Voter's Guide to Election Polls*, 94.

38. Bowman, "Polluted Polling on Global Warming."

39. Gerstenzang, "Survey Bolsters Global Warming Fight," A4.

40. In this case, the bias of the World Wildlife Fund question seems particularly unfortunate since many other survey organizations obtained similar results without risking credibility. Seventy-seven percent of respondents in a Ohio State University survey funded by the National Science Foundation believed global warming "has been happening," while 74 percent believed that global warming was "real" in a Wirthlin Worldwide (1997) survey.

41. Immerwahr, *Waiting for a Signal.*

42. See, for example, Dunlap, "Public Opinion and Environmental Policy," in Lester, ed., *Environmental Politics and Policy*; he argues that the direction and degree of public opinion are the "most basic concepts of this kind of study" (p. 123).

42. Key, *Public Opinion and American Democracy.*

Chapter 2

1. In abbreviated form, the work of both Bryce and Lowell can be found in Janowitz and Hirsch, eds., *Reader in Public Opinion and Mass Communication*. For the full work of each, see Bryce, *The American Commonwealth*, Vol. II, and Lowell, *Public Opinion and Popular Government.*

2. Key, *Public Opinion and American Democracy.*

3. Dunlap, "Public Opinion in the 1980s"; see also Shabecoff, *Earth Rising.*

4. Dunlap and Scarce, "The Polls"; also Dunlap, "Public Opinion and Environmental Policy," in Lester, ed., *Environmental Politics and Policy.*

5. Schuman and Presser, *Questions and Answers in Attitude Surveys.*

6. The "filter" described by Sterngold, Warland, and Herrmann in "Do Surveys Overstate Public Concerns?" would first ask for an initial expression of concern, pressing forward with additional measures of intensity only if respondents passed that litmus test—in essence, separating the measurement of direction from intensity rather than combining both into a single response format. Given constraints of time and expense, however, this safeguard is not always adopted in practice (and is, in fact, not used by Gallup in the poll discussed here). For similar views on the importance of filter questions, see Key, *Public Opinion*, chapter 9, and Schuman and Presser, *Questions and Answers*, chapters 9 and 10.

7. Ladd and Bowman, *Attitudes toward the Environment*, 17, 20.

8. In "Public Opinion," Dunlap argues that "ranking procedures" can be reliable indictors of attitude strength and intensity. In Lester, ed., *Environmental Politics and Policy.*

9. Traugott and Lavrakas, *The Voter's Guide to Election Polls.*

10. Wirthlin Worldwide National Quorum Survey (1998), Q9.

11. Even though extremity of opinion and intensity of opinion are related (see Schuman and Presser, *Questions and Answers*, 233).

12. Including Dunlap, "Public Opinion," in Lester, ed., *Environmental Politics and Policy*, 124–125.

13. Funkhouser, "The Issues of the Sixties"; Dunlap, "Trends in Public Opinion Toward Environmental Issues."

14. Mitchell, "Public Opinion and Environmental Politics in the 1970s and 1980s," in Vig and Kraft, eds., *Environmental Policy in the 1980s.*

15. Dunlap, "Public Opinion," in Lester, ed., *Environmental Politics and Policy*, 128.

16. Dunlap and Scarce, "The Polls," 653.

17. Ladd and Bowman, *Attitudes toward the Environment*, 20.

18. Mitchell, "Public Opinion and Environmental Politics in the 1970s and 1980s," in Vig and Kraft, eds., *Environmental Policy in the 1980s.*

19. Respondents who have little strength of feeling for the environment are presumably those most likely to be influenced by counterarguments because they have little commitment to the position in the first place. See Mitchell, "Silent Spring/Solid Majorities" and *Public Opinion on Environmental Issues*; also Keeter, "Problematical Pollution Polls."

20. The Gallup Organization [datafile], April 3–9, 2000 ($n = 1,004$ adults nationwide). Margin of error ± 3 percentage points.

21. Ladd and Bowman, *Attitudes toward the Environment*, 24; see also Ladd, "What Do Americans Really Think About the Environment?"

22. Cambridge Reports National Omnibus Survey [datafile], September 1994 ($n = 1,250$ adults nationwide): Q11, Q13.

23. Ladd, "Clearing the Air," 19. See also Keeter (1984).

24. Keeter, "Problematical Pollution Polls," 274.

25. Ladd and Bowman, *Attitudes toward the Environment*, 24.

26. Schuman and Presser, *Questions and Answers*, 298–299.

27. See, for example, Schuman and Presser, *Questions and Answers*, 173–174.

28. Boudreaux, Meiners, and Zywicki, "Talk Is Cheap," 765.

29. Smith, "Can We Measure the Economic Value of Environmental Amenities?"; Hotelling, "Letter to Director of National Park Service," in *An Economic Study of the Monetary Evaluation of Recreation in the National Parks*; Dwyer, "Economic Benefits of Wildlife-Related Recreation Experiences," in Shaw and Zube, eds., *Wildlife Values*; Schechter, Kim, and Golan, "Valuing a Public Good," in Folmer, and ver Ierland, eds., *Valuation Methods and Policy Making in Environmental Economics.*

30. For example, an individual may value the option of using an environmental resource, such as a national park, at some time in the future above and beyond how likely they are to use it and thus be willing to pay under conditions of uncertainty to ensure it will be kept available and unspoiled. In addition, an individual may value and derive benefit from the mere existence of an environmental resource, such as the survival of a particular species of wildlife or the natural condition of the Grand Canyon. See Bishop, "Option Value"; Weisbrod, "Collective-Consumption Services of Individual-Consumption Goods"; Krutilla, "Conservation Reconsidered"; Krutilla and Fisher, *The Economics of Natural Environments*.

31. Mitchell and Carson, *Using Surveys to Value Public Goods*; Smith, "Can We Measure the Economic Value of Environmental Amenities?"

32. The 72 percent cited combines two response categories, "very willing" and "somewhat willing." Likewise, the NORC GSS numbers combine similar answers, "very willing" and "fairly willing." Louis Harris and Associates (1993), Q4; NORC General Social Survey (1994).

33. Cambridge Reports/Research International (July 1994), Q9; Yankelovich Clancy Shulman (1992), Q7.

34. Princeton Survey Research Associates (1997), Q39; Pew Research Center (1997), Q75.

35. Boudreaux, Meiners, and Zywicki, "Talk Is Cheap"; Fischer, "Willingness to Pay as a Behavioral Criterion for Environmental Decision-making."

36. Green, Kahneman, and Kunreuther, "How the Scope and Method of Public Funding Affect Willingness to Pay for Public Goods," 49.

37. For example, as Sunstein notes in "Experts, Economists, and Democrats," willingness-to-pay to protect spotted owls "drops significantly" when the wording of the question values the owl with and in comparison to other species, rather than in isolation; see also Fischer, "Willingness to Pay."

38. Ladd, "Clearing the Air," 11. But, as he argues, this is not different from other issues, including health care for the uninsured. In general, Americans are not willing to pay even small amounts for some things they care deeply about.

39. This issue, of course, is not new. Most work on voter turnout and vote choice involves the use of self-reported behavior, which leads to problems of overreporting. See Katosh and Traugott, "The Consequences of Validated and Self-Reported Voting Measures"; Silver, Anderson, and Abramson, "Who Overreports Voting?"

40. Presser, "Can Changes in Context Reduce Vote Overreporting in Surveys?"

41. Furman and Erdur, "Are Good Citizens Environmentalists?"

42. Dunlap and Scarce, "The Polls," 657.

43. Oskamp, Harrington, Edwards, Sherwood, Okuda, and Swanson, "Factors Influencing Household Recycling Behavior."

44. Milbrath, *Environmentalists*; also Dunlap, "Public Opinion in the 1980s."

45. De Young, "Encouraging Environmentally Appropriate Behavior" and "Some Psychological Aspects of Recycling."

46. Scott Geller, Winett, and Everett, *Preserving the Environment.*

47. Ladd and Bowman, *Attitudes toward the Environment.*

48. Ayres, "The Worldwatch Report," available at ⟨http://www.enn.com/enn-features-archive/1999/05/051299/worldwatch_3141.asp⟩ [emphasis added].

49. Mitchell, "Public Opinion and Environmental Politics"; Dunlap, "Public Opinion." See also Dunlap and Dillman, "Decline in Public Support for Environmental Protection."

50. For example only 14 percent of those responding in a 1990 Gallup poll believed that "a great deal" of progress had been made on environmental issues, while ten years later that number had grown to 26 percent. In both cases, however, a majority (63 percent and 64 percent, respectively) thought that "only some progress" had been achieved. See the Gallup Organization (1990), Q18, and (2000), Q43.

51. The Gallup Organization (2000), Q44.

52. Mitchell, "Public Opinion and Environmental Politics," 55.

53. Immerwahr, *Waiting for a Signal.*

54. Baldassare and Katz, "The Personal Threat of Environmental Problems as Predictor of Environmental Practices"; Rohrschneider, "Citizens' Attitudes toward Environmental Issues."

55. Shabecoff, *Earth Rising*, 38.

56. Rohrschneider, "Citizens' Attitudes toward Environmental Issues."

57. Helvarg, *The War Against the Greens.*

58. Kempton, Boster, and Hartley, *Environmental Values in American Culture.*

59. Dunlap, "Public Opinion," 12, 32.

60. For example, in the April 2000 Gallup poll cited previously in this chapter, respondents were asked: "Thinking specifically about the environmental movement, do you think of yourself as an active participant in the environmental movement [16 percent], sympathetic toward the movement, but not active [55 percent], neutral [23 percent], or unsympathetic toward the environmental movement [5 percent]?"

61. Dowie, "American Environmentalism."

Chapter 3

1. Downs, "Up and Down with Ecology," 38.

2. Downs, "Up and Down with Ecology," 39; see also Albrecht and Mauss, "The Environment as a Social Problem," in Mauss, ed., *Social Problems as Social Movements.*

3. Downs, "Up and Down with Ecology," 50.

4. Erskine, "The Polls," 120.

5. See, for example, Hornback, "Orbits of Opinion"; Buttel, "Class Conflict, Environmental Conflict, and the Environmental Movement"; Dunlap and Dillman, "Decline in Public Support for Environmental Protection"; Dunlap and Van Liere, "Further Evidence of Declining Public Concern with Environmental Problems"; Marsh and Christenson, "Support for Economic Growth and Environmental Protection"; Rosenbaum, *The Politics of Environmental Concern*; Dunlap, Van Liere, and Dillman, "Evidence of Decline in Public Concern with Environmental Quality"; Dunlap, "Public Opinion and Environmental Policy," in Lester, ed., *Environmental Politics and Policy*; Dunlap, "Trends in Public Opinion Toward Environmental Issues."

6. Dunlap and Scarce, "The Polls," 652.

7. Stisser, "A Deeper Shade of Green," 24.

8. Dunlap and Dillman, "Decline in Public Support for Environmental Protection"; Marsh and Christenson, "Support for Economic Growth and Environmental Protection"; see also Albrecht and Mauss, "The Environment as a Social Problem."

9. Key, "*Public Opinion and American Democracy*," 234.

10. Fishbein and Ajzen. 1975. *Belief, Attitude, Intention, and Behavior*, 6.

11. For a prime example, see Converse, "The Nature of Belief Systems in Mass Publics," in Apter, ed., *Ideology and Discontent*; see also Smith, "Is There Real Opinion Change?"

12. Smith, "Is There Real Opinion Change?"; Page and Shapiro, "Changes in Americans' Policy Preferences."

13. Note that in the absence of panel studies, aggregate trends remain the only viable possibility for judging long-term attitude change.

14. Glenn, "Trend Studies with Available Survey Data," in *Survey Data for Trend Analysis*.

15. Taylor, "Procedures for Evaluating Trends in Public Opinion." See also Smith, "A Compendium of Trends on General Social Survey Questions."

16. See Smith, "Is There Real Opinion Change?" One well-known example of a "constant" model suits Sullivan et al.'s data demonstrating that levels of political tolerance in the United States have not changed in recent years. See Sullivan, Piereson, and Marcus, *Political Tolerance and American Democracy*.

17. Smith, "Cycles of Reform?"

18. This type of change is sometimes referred to as *bounce*. Smith, in "Is There Real Opinion Change?," notes that studies of presidential popularity often fit this mold.

19. On the importance of birth cohorts in understanding environmental attitudes, see Hays, *Beauty, Health, and Permanence*; Mohai and Twight, "Age and

Environmentalism"; Buttel, "Age and Environmental Concern"; Honnold, "Age and Environmental Concern."

20. Dunlap, "Trends in Public Opinion toward Environmental Issues."

21. For a detailed description of trends, see Dunlap, "Trends in Public Opinion toward Environmental Issues"; Dunlap and Scarce, "The Polls"; also Gillroy and Shapiro, "The Polls."

22. For a full summary of that literature, refer to Dunlap, "Trends in Public Opinion."

23. This result confirms Tom Smith's conclusion regarding the same variable between 1972 and 1977. See Smith, "Age and Social Change." To put that figure in some perspective, Smith finds in his study of 455 liberal/conservative measures from 1936 to 1985 that linear trends across many different variables averaged just 1.3 percentage points per annum. See Smith, "Liberal and Conservative Trends in the United States since World War II."

24. Bear in mind that the proportion of variance in attitudes explained by the equation is quite small (R-square = .175), which means that the bulk—nearly 83 percent—remains unexplained using a purely linear model.

25. Key, "Public Opinion and American Democracy," 235.

26. For a more detailed explanation, refer to Firebaugh and Davis, "Trends in Anti-Black Prejudice, 1972–1984."

27. Kanagy, Humphrey, and Firebaugh, "Surging Environmentalism."

28. For example, see Durr, "What Moves Policy Sentiment?"

29. Albrecht and Mauss, "The Environment as a Social Problem"; Dunlap and Dillman, "Decline in Public Support for Environmental Protection"; Marsh and Christenson, "Support for Economic Growth and Environmental Protection."

30. Alexander, "Gunning for the Greens," 50.

31. Lauter and Jehl, "Bush, Clinton Clash Over Jobs, Environment," A1.

32. For an intriguing account of the gap between the reality and perception of economic conditions during the 1992 presidential election, see Hetherington, "The Media's Role in Forming Voters' National Economic Evaluations in 1992."

33. Elliott, Regens, and Seldon, "Exploring Variation in Public Support for Environmental Protection," 51; also Elliott, Seldon, and Regens, "Political and Economic Determinants of Individuals' Support for Environmental Spending."

34. See, for example, Dunlap, "Public Opinion and Environmental Policy," 90; and Dunlap and Dillman, "Decline in Public Support for Environmental Protection."

35. Wlezien, "The Public as Thermostat," 981.

36. Karlberg, "News and Conflict"; also MacKuen, "Reality, the Press, and Citizens' Political Agendas," in Turner and Martin, eds., *Surveying Subjective Phenomena*.

37. Page, Shapiro, and Dempsey, "What Moves Public Opinion?," 24.

38. Allan, Adam, and Carter, "Introduction," in *Environmental Risks and the Media*, 4. See also Parlour and Schatzow, "The Mass Media and Public Concern for Environmental Problems in Canada, 1960–1972."

39. Missing values were interpolated in SAS using a cubic spline curve to create an uninterrupted time-series.

40. Readers unfamiliar with basic regression techniques may find the following helpful: Berry and Sanders, *Understanding Multivariate Research*.

41. See Ostrom, *Time Series Analysis*; also Kennedy, *A Guide to Econometrics*, chapter 8; and Choudhury, Hubata, and St. Louis, "Understanding Time-Series Regression Estimators."

42. King, *Unifying Political Methodology*, 181.

43. Durr, "What Moves Policy Sentiment?," 164.

44. The Durbin-Watson statistic tests for the absence of first-order autocorrelation in ordinary least squares (OLS) residuals. With possible values ranging from 0 to 4, Durbin-Watson statistics closest to a midpoint of 2 (with probabilities that fall outside the conventional range of statistical significance) allow researchers to accept the null hypothesis that there is no significant correlation. In this case the Durbin-Watson of the model presented equaled 2.06 with a probability of 0.13.

45. For example, Kanagy, Humphrey, and Firebaugh, in "Surging Environmentalism," used pooled cross-sectional data to isolate birth cohorts from period effects.

46. This same difficulty in finding an appropriations measure commensurate with the questions used in public opinion polls is noted by Wlezien, "The Public as Thermostat," 987.

47. For a more detailed description of the PACE index, see Streitwieser, "Using the Pollution Abatement Costs and Expenditures Micro Data for Descriptive and Analytic Research."

48. Environmental Protection Agency, *Environmental Investments*. See also Carlin, Scodari, and Garner, "Environmental Investments."

49. This result is consistent with Robert Durr's contention that an "optimistic economic outlook results in greater support for liberal domestic policies." See Durr, "What Moves Policy Sentiment?," 167.

50. See, for example, Page and Shapiro, "Effects of Public Opinion on Policy."

51. Wlezien, "The Public as Thermostat," 993.

52. For example, between 1973 and 1976, federal budgets under Republican president Richard M. Nixon appropriated (on average) 2.12 percent of total budget dollars to natural resources and the environment, followed by 2.40 percent under Democrat Jimmy Carter, 1.67 percent under Republican Ronald Reagan, 1.39 percent under Republican George H. W. Bush, and 1.41 percent under Democrat Bill Clinton.

53. Media attention alone does not always produce increased environmental concern. As Daley and O'Neill argue in "Sad Is Too Mild a Word," content matters. In news coverage of the *Exxon Valdez* oil spill, for example, they argue that the news media "naturalised" the disaster, directing attention "away from the political arena and into the politically inaccessible realm of technological inevitability" (p. 53). For another pair of contrasting views on the accuracy of news reporting on the environment, refer to Ruben, "Back Talk"; and Michaels, "Reporters Cry Wolf about Environment."

54. Smith, "Is There Real Opinion Change?"

55. Dunlap, "Trends in Public Opinion toward Environmental Issues."

56. Stimson, *Public Opinion in America.*

57. For additional confirmation of the stable priority accorded to environmental protection despite the competition of other issues, refer to Lowe, Pinhey, and Grimes, "Public Support for Environmental Protection."

58. Kingdon, *Agendas, Alternatives, and Public Policies.*

59. For example, "One might conclude ... that efforts at initiating liberal policies are ill advised during troubled economic times" and that external factors like the economy "play a critical role in opening and closing the 'policy windows' through which the makers and advocates of policy initiatives must move." See Durr, "What Moves Policy Sentiment?," 167.

Chapter 4

1. Anthony, "Race, Justice, and Sprawl," 97.

2. Bullard, *Dumping in Dixie.*

3. Easterbrook, "Suburban Myth," 18.

4. Loth, "Bringing Earth Day Back Down to Earth," A33.

5. Dowie, *Losing Ground*, 6.

6. Loth, "Bringing Earth Day Back Down to Earth," A33.

7. Sagoff, "The Great Environmental Awakening," 46.

8. Constantini and Hanf, "Environmental Concern and Lake Tahoe"; Dillman and Christenson, "The Public Value for Pollution Control," in Burch, Cheek, and Taylor, eds., *Social Behavior, Natural Resources, and the Environment*; Tognacci, Weigel, Wideen, and Vernon, "Environmental Quality"; Albrecht and Mauss, "The Environment as a Social Problem," in Mauss, ed., *Social Problems as Social Movements*; Dunlap, "The Impact of Political Orientation on Environmental Attitudes and Action."

9. Ogden, "The Future of the Environmental Struggle," in Meek and Straayer, eds., *The Politics of Neglect*; Mitchell, "Silent Spring/Solid Majorities"; Lowe, Pinhey, and Grimes, "Public Support for Environmental Protection"; Ladd, "Clearing the Air"; Mohai, "Public Concern and Elite Involvement in Environ-

mental-Conservation Issues"; Samdahl and Robertson, "Social Determinants of Environmental Concern"; Dietz, Stern, and Guagnano, "Social Structural and Social Psychological Bases of Environmental Concern."

10. For an excellent argument on this topic, see Klineberg, McKeever, and Rothenbach, "Demographic Predictors of Environmental Concern."

11. Dietz, Stern, and Guagnano, "Social Structural and Social Psychological Bases of Environmental Concern," 463.

12. Denton Morrison and Riley Dunlap make a critical distinction between "compositional elitism," which suggests that environmentalists are drawn from an upper socioeconomic status; "ideological elitism," which argues that environmental policy agendas benefit environmentalists at the expense of other groups; and "impact elitism," where environmental policies have regressive social consequences. The specific charge of compositional elitism is what is explored here. See Morrison and Dunlap, "Environmentalism and Elitism."

13. Readers unfamiliar with regression analysis might find the analysis in this chapter confusing at times. A basic primer should help, such as Berry and Sanders, *Understanding Multivariate Research.*

14. Mohai and Twight, "Age and Environmentalism."

15. In other words, the effect of age relates to birth cohorts and not to simple biological maturity. See Hays, *Beauty, Health, and Permanence*; Malkis and Grasmick, "Support for the Ideology of the Environmental Movement." On more general theories regarding the importance of birth cohorts, see Ryder, "The Cohort as a Concept in the Study of Social Change"; Schuman and Rieger, "Historical Analogies, Generational Effects, and Attitudes toward War."

16. Inglehart, "Post-Materialism in an Environment of Insecurity," 882.

17. It is difficult, however, to distinguish between cohort and period effects within a single cross-sectional survey. One recent study that attempts to do just that argues that specific historical triggers—those that influence birth cohorts in equal measure, including the Chernobyl nuclear disaster and the *Exxon Valdez* oil spill in Alaska—are more powerful in the end. See Kanagy, Humphrey, and Firebaugh, "Surging Environmentalism."

18. Munton and Brady, "American Public Opinion and Environmental Pollution," 18; Van Liere and Dunlap, "The Social Bases of Environmental Concern"; Albrecht and Mauss, "The Environment as a Social Problem," 578.

19. Verba and Nie, *Participation in America*; Milbrath and Goel, *Political Participation*; Conway, *Political Participation in the United States*; Brady, Verba, and Scholzman, "Beyond SES."

20. See, for example, Gottlieb, "Forcing the Spring"; and Paehlke, *Environmentalism and the Future of Progressive Politics.*

21. Douglas and Wildavsky, *Risk and Culture*; also, Paehlke, *Environmentalism and the Future of Progressive Politics.*

22. As Van Liere and Dunlap reason in "The Social Bases of Environmental Concern," "Members of the lower class typically have experienced only poor physical conditions, and thus are less aware that they live, work, and play in polluted, overcrowded conditions" (p. 184). These theories are often extended to comparisons between developed and developing countries. See Dunlap and Mertig, "Global Concern for the Environment"; also Diekmann and Franzen, "The Wealth of Nations and Environmental Concern."

23. Loth, "Bringing Earth Day Back Down to Earth," A33.

24. Richard Rodriguez, quoted in Dowie, *Losing Ground*, 220. See also Hershey and Hill, "Is Pollution a 'White Thing?'"; Taylor, "Blacks and the Environment"; Baugh, "African-Americans and the Environment."

25. As others point out, however, environmental health threats (air pollution, in particular) are often most severe for these groups. See Bullard, *Dumping in Dixie*.

26. Bullard, *Dumping in Dixie*, xiv–xv.

27. For a review of that field of research, see Davidson and Freudenburg, "Gender and Environmental Risk Concerns"; also Zelezny, Chua, and Aldrich, "Elaborating on Gender Differences in Environmentalism."

28. Mohai, "Men, Women, and the Environment"; also Blocker and Eckberg, "Gender and Environmentalism"; Gottlieb, "Forcing the Spring," chapter 6.

29. Bord and O'Connor, "The Gender Gap in Environmental Attitudes."

30. Davidson and Freudenburg, "Gender and Environmental Risk Concerns."

31. See Mohai, "Men, Women, and the Environment"; also Nelkins, "Nuclear Power as a Feminist Issue"; Brody, "Differences by Sex in Support for Nuclear Power"; Blocker and Eckberg, "Environmental Issues as Women's Issues"; Solomon, Tomaskovic-Devey, and Risman, "The Gender Gap and Nuclear Power"; also Blocker and Eckberg, "Gender and Environmentalism."

32. The nature of that criticism is outlined by Sagoff, in "The Great Environmental Awakening," but not ultimately supported by him.

33. Dietz, Stern, and Guagnano, "Social Structural and Social Psychological Bases of Environmental Concern." See also Buttel and Flinn, "The Politics of Environmental Concern"; also Constantini and Hanf, "Environmental Concern and Lake Tahoe"; Dillman and Christenson, "The Public Value for Pollution Control"; Tognacci et al., "Environmental Quality"; Dunlap, "The Impact of Political Orientation on Environmental Attitudes and Action"; Van Liere and Dunlap, "The Social Bases of Environmental Concern"; and Samdahl and Robertson, "Determinants of Environmental Concern."

34. Van Liere and Dunlap, "The Social Bases of Environmental Concern," 192.

35. See, for example, Dietz, Stern, and Guagnano, "Social Structural and Social Psychological Bases of Environmental Concern"; also Klineberg, McKeever, and Rothenbach, "Demographic Predictors of Environmental Concern." The effect of race on environmental attitudes is questioned in at least two recent studies:

Mohai and Bryant, "Is There a 'Race' Effect on Concern for Environmental Quality?"; and Jones and Carter, "Concern for the Environment among Black Americans." The influence of gender, too, has been refuted by some, including Dietz, Stern, and Guagnano, "Social Structural and Social Psychological Bases of Environmental Concern"; and Blocker and Eckberg, "Environmental Issues as Women's Issues."

36. Klineberg, McKeever, and Rothenbach, "Demographic Predictors of Environmental Concern," 749.

37. Buttel and Flinn, "The Politics of Environmental Concern"; Weigel and Weigel, "Environmental Concern."

38. Dillman and Christenson, "The Public Value for Pollution Control"; Elliot, Regens, and Seldon, "Exploring Variation in Public Support for Environmental Protection"; Elliott, Seldon, and Regens, "Political and Economic Determinants of Individuals' Support for Environmental Spending."

39. Maloney, Ward, and Braucht, "A Revised Scale for the Measurement of Ecological Attitudes and Knowledge."

40. Buttel and Flinn, "The Politics of Environmental Concern."

41. Dunlap, Grieneeks, and Rokeach, "Human Values and Pro-Environmental Behavior," in Conn, ed., *Energy and Material Resources*; Milbrath, *Environmentalists*; Schwartz and Miller, "The Earth's Best Friends"; Mohai, "Public Concern and Elite Involvement in Environmental-Conservation Issues."

42. For example, there is much evidence to suggest that relatively affluent, better-educated individuals are more politically active in all issue areas. See Milbrath, *Political Participation*; also Verba and Nie, *Participation in America*.

43. Klineberg, McKeever, and Rothenbach, "Demographic Predictors of Environmental Concern," 749; see also Van Liere and Dunlap, "Environmental Concern"; and Samdahl and Robertson, "Social Determinants of Environmental Concern."

44. It should be noted that NORC has not repeated its battery of General Social Survey (GSS) environmental questions since 1994. The data used here, therefore, are the most recent available.

45. Berry and Sanders, *Understanding Multivariate Research*, 45–49.

46. Douglas and Wildavsky, *Risk and Culture*.

47. To isolate this possibility more directly, data were broken down by residency, including urban, suburban, small-town, and rural areas, but no independent geographic effects were found. This suggests that income differences could also be the result of a gap between subjective and objective reality, where low-income respondents view environmental risks in personal terms irrespective of their surroundings. In this case, the survey context is limiting in that low-income respondents cannot be observed under different known environmental conditions.

48. Bullard, *Dumping in Dixie*.

49. The R-square of a regression equation measures the explanatory power (or "fit") of the model. Ranging in value from 0 to 1, it represents the proportion of the variation in the dependent variable (y) that is explained by the independent variables ($X_1 \ldots X_n$) in the model. See Berry and Sanders, *Understanding Multivariate Research*, 44–45.

50. Ordered probit is necessary here because of the nature of the dependent variables used. Ordinary least squares regression (OLS) assumes under the Gauss-Markov theorem that the dependent variable is continuous and is free to take on any value from negative infinity to positive infinity. While the accuracy of this assumption can be questioned on many measures used in survey research, it is especially violated in cases where responses are coded into a small number of response categories. In the 1994 NORC General Social Survey (GSS), for example, respondents are asked to consider whether spending on the environment is currently "too little," "about the right amount," or "too much," coded as 1, 2, and 3, respectively. In this case, since the dependent variable is ordinal, OLS no longer provides the best linear unbiased estimator. Indeed, it can be shown that OLS estimates are both biased and inefficient and that they no longer have the smallest possible sampling variance. To correct for the use of a limited dependent variable, a nonlinear probability model, such as ordered probit, should be used to fit the data. While the underlying logic of this approach is much the same as the logic of approaches used in more basic regression models, slope coefficients produced using probit cannot be interpreted simply as the marginal effect of an independent variable. Some additional calculation is required. Readers unfamiliar with the technique should refer to Aldrich and Nelson, *Linear Probability, Logit, and Probit Models*.

51. In other words, "correlations with political ideology may be less apparent when environmental concern is measured in other ways." See Klineberg, McKeever, and Rothenbach, "Demographic Predictors of Environmental Concern," 737.

52. Berelson, Lazarsfeld, and McPhee, *Voting*; also Campbell, Converse, Miller, and Stokes, *The American Voter*.

53. Campbell et al., *The American Voter*, 296.

54. Mohai, "Black Environmentalism."

55. Those arguments are summarized but ultimately refuted by Sagoff in "The Great Environmental Awakening," 44, 46.

56. Tucker, *Progress and Privilege*, 34–36; Douglas and Wildavsky, *Risk and Culture*; Sagoff, "The Great Environmental Awakening," 40.

57. Schneider, "Everybody's an Environmentalist Now," 1062.

58. See, for example, Constantini and Hanf, "Environmental Concern and Lake Tahoe"; Dillman and Christenson, "The Public Value for Pollution Control"; Tognacci et al., "Environmental Quality"; Albrecht and Mauss, "The Environment as a Social Problem"; Dunlap, "The Impact of Political Orientation on Environmental Attitudes and Action."

59. That conclusion is corroborated by Robert Emmet Jones and Riley E. Dunlap in "The Social Bases of Environmental Concern"; also, Klineberg, McKeever, and Rothenbach, "Demographic Predictors of Environmental Concern."

60. Gottlieb, "Forcing the Spring," 306.

61. See, for example, Kempton, Boster, and Hartley, *Environmental Values in American Culture*.

Chapter 5

1. Converse, "The Nature of Belief Systems in Mass Publics," in Apter, ed., *Ideology and Discontent*.

2. The notion of attitude consistency (or "constraint") was central to Converse's thesis, which defined a belief system as a "configuration of ideas and attitudes in which the elements are bound together by some form of constraint or functional interdependence." See, for example, Converse, "The Nature of Belief Systems," 207.

3. Converse, "The Nature of Belief Systems," 245.

4. Zaller and Feldman, "A Simple Theory of the Survey Response," 580.

5. Zaller and Feldman, "A Simple Theory." For results similar to Converse, "The Nature of Belief Systems," refer to Campbell, Converse, Miller, and Stokes, *The American Voter*; also McCloskey, "Consensus and Ideology in American Politics."

6. Key, "Public Opinion and American Democracy," 153. For more on the importance of cognitive consistency theory, see Bennett, "Consistency among the Public's Social Welfare Policy Attitudes in the 1960s."

7. Swan, "Environmental Education."

8. Rosenbaum, *Environmental Politics and Policy*, 305.

9. For further discussion of that "gap," see Dunlap, "Polls, Pollution, and Politics Revisited"; "Public Opinion and Environmental Policy" in Lester, ed., *Environmental Politics and Policy*; and "Public Opinion in the 1980s"; also Schwartz and Miller, "The Earth's Best Friends."

10. For a deeper discussion of "doorstep opinions," see Erikson, Luttbeg, and Tedin, *American Public Opinion*.

11. deHaven-Smith, "Toward a Communicative Theory of Environmental Opinion," 630.

12. To restate the point more formally, the presence of non-random measurement error means that correlations between latent traits of interest may bear little resemblance to those between observed indicators. See Blalock, "Some Implications of Random Measurement Error for Causal Inferences," "Multiple Indicators and the Causal Approach to Measurement Error," and "A Causal Approach to Nonrandom Measurement Errors"; Achen, "Mass Political Attitudes and the

Survey Response"; Green, "On the Dimensionality of Public Sentiment toward Partisan and Ideological Groups"; Green and Citrin, "Measurement Error and the Structure of Attitudes."

13. Tognacci et al., "Environmental Quality."

14. Dillman and Christenson, "The Public Value for Pollution Control," in Burch, Cheek, and Taylor, eds., *Social Behavior, Natural Resources, and the Environment*; Marsh and Christenson, "Support for Economic Growth and Environmental Protection, 1973–1975."

15. Maloney, Ward, and Braucht, "A Revised Scale for the Measurement of Ecological Attitudes and Knowledge."

16. Buttel and Flinn, "The Politics of Environmental Concern"; Buttel and Johnson, "Dimensions of Environmental Concern."

17. Dunlap, Grieneeks, and Rokeach, "Human Values and Pro-Environmental Behavior," in Conn, ed., *Energy and Material Resources*; Milbrath, *Environmentalists*; Schwartz and Miller, "The Earth's Best Friends."

18. Buttel and Flinn, "The Politics of Environmental Concern"; Maloney, Ward, and Braucht, "A Revised Scale for the Measurement of Ecological Attitudes and Knowledge."

19. Van Liere and Dunlap, "Environmental Concern," 652.

20. Most recently, see Klineberg, McKeever, and Rothenbach, "Demographic Predictors of Environmental Concern," 734.

21. On the weak relationship between various measures of environmental attitudes, see Lounsbury and Tornatzky, "A Scale for Assessing Attitudes toward Environmental Quality"; Albrecht, Bultena, Hoiberg, and Nowak, "The New Environmental Paradigm Scale"; Keeter, "Problematical Pollution Polls"; Geller and Lasley, "The New Environmental Paradigm Scale"; also Buttel and Johnson, "Dimensions of Environmental Concern"; Van Liere and Dunlap, "Environmental Concern." It is interesting to note that evidence also suggests that pro-environmental behaviors, such as recycling, carpooling, and energy conservation, are not strongly correlated. For a review of that literature, refer to Mainieri, Barnett, Valdero, Unipan, and Oskamp, "Green Buying."

22. Tognacci et al., "Environmental Quality."

23. Buttel and Johnson, "Dimensions of Environmental Concern."

24. Dunlap and Van Liere, "The 'New Environmental Paradigm'" and "Commitment to the Dominant Social Paradigm and Concern for Environmental Quality."

25. Pirages and Ehrlich, *Ark II*; Catton and Dunlap, "Environmental Sociology"; also Dunlap and Van Liere, "The 'New Environmental Paradigm'"; Milbrath, *Environmentalists*.

26. Dunlap and Van Liere's original battery of survey questions included the following twelve statements:

• We are approaching the limit of the number of people the earth can support.
• The balance of nature is very delicate and easily upset.
• Humans have the right to modify the natural environment.
• Humankind was created to rule over the rest of nature.
• When humans interfere with nature it often produces disastrous consequences.
• Plants and animals exist primarily to be used by humans.
• To maintain a healthy economy we will have to develop a steady-state economy where industrial growth is controlled.
• Humans must live in harmony with nature in order to survive.
• The earth is like a spaceship with only limited room and resources.
• Humans need not adapt to the natural environment because they can remake it to suit their needs.
• There are limits to growth beyond which our industrialized society cannot expand.
• Mankind is severely abusing the environment.

See Dunlap and Van Liere, "The 'New Environmental Paradigm.'"

27. Albrecht et al., "The New Environmental Paradigm Scale."

28. Geller and Lasley, "The New Environmental Paradigm Scale."

29. Pierce, Lovrich, and Tsurutani, "Culture, Politics and Mass Publics"; Kuhn and Jackson, "Stability of Factor Structures in the Measurement of Public Environmental Attitudes"; Noe and Snow, "The New Environmental Paradigm and Further Scale Analysis"; Scott and Willits, "Environmental Attitudes and Behavior"; Roberts and Bacon, "Exploring the Subtle Relationships between Environmental Concern and Ecologically Conscious Consumer Behavior."

30. Dunlap, Van Liere, Mertig, and Jones, "Measuring Endorsement of the New Ecological Paradigm."

31. deHaven-Smith, "Toward a Communicative Theory," 194.

32. For example, Dunlap and Van Liere, "The 'New Environmental Paradigm'"; Van Liere and Dunlap, "Environmental Concern"; Kuhn and Jackson, "Stability of Factor Structures in the Measurement of Public Environmental Attitudes"; Scott and Willits, "Environmental Attitudes and Behavior."

33. Green and Citrin, "Measurement Error and the Structure of Attitudes," 261fn.

34. See, for example, Dunlap and Van Liere, "The 'New Environmental Paradigm'"; Geller and Lasley, "The New Environmental Paradigm Scale."

35. Green and Citrin, "Measurement Error and the Structure of Attitudes."

36. For advanced readers, a brief mathematical explanation of measurement error (and its effects) is offered here. Let

$X_{1i} = \xi_i + \delta_{1i}$

$X_{2i} = \lambda\xi_i + \delta_{2i}$

where

X_{ki} represents the observed variables,
ξ_i is the underlying trait,
δ_{1i} represents errors in measurement, and
λ is a scaling coefficient.

Example 1: No measurement error If no measurement error is present,
$$\text{VAR}(\delta_{1i}) = \text{VAR}(\delta_{1i}) = 0$$
Correlation between variables is calculated using the following equation:
$$r(X_{1i}, X_{2i}) = \frac{\text{COV}(X_{1i}, X_{2i})}{\sqrt{\text{VAR}(X_{1i})} \cdot \sqrt{\text{VAR}(X_{2i})}} = \frac{\lambda\,\text{VAR}(\xi_i)}{\sqrt{\text{VAR}(\xi_i)} \cdot \sqrt{\lambda^2\,\text{VAR}(\xi_i)}}$$

Example 2: Random measurement error only If random errors in measurement occur,
$$\left[\begin{array}{l} \text{where, VAR}(\delta_{1i}) \text{ and VAR}(\delta_{2i}) > 0, \\ \text{but COV}(\xi_i, \delta_{ki}) = \text{COV}(\delta_{1i}, \delta_{2i}) = 0 \end{array}\right]$$
$$r(X_{1i}, X_{2i}) = \frac{\lambda\,\text{VAR}(\xi_i)}{\sqrt{\text{VAR}(\xi_i) + \text{VAR}(\delta_{1i})} \cdot \sqrt{\lambda^2\,\text{VAR}(\xi_i) + \text{VAR}(\delta_{2i})}}$$

Note, because of additional terms in the denominator, the correlation between variables in this example is weaker than it is in example 1. In other words, correlations contaminated with random measurement error will be attenuated.

Example 3: Random and nonrandom measurement error
$$[\text{where, COV}(\delta_{1i}, \delta_{2i}) \neq 0]$$
$$r(X_{1i}, X_{2i}) = \frac{\lambda\,\text{VAR}(\xi_i) + \text{COV}(\delta_{1i}, \delta_{2i})}{\sqrt{\text{VAR}(\xi_i) + \text{VAR}(\delta_{1i})} \cdot \sqrt{\lambda^2\,\text{VAR}(\xi_i) + \text{VAR}(\delta_{2i})}}$$

When nonrandom measurement error is also present, note that the effect on r is less predictable. The correlation in example 3 could be either larger *or* smaller than the correlation in example 1 and could even be of the wrong sign, depending on the relative size of the error variance and covariance terms. For more background on measurement error and its effects, refer to Blalock, "Some Implications of Random Measurement Error," "Multiple Indicators and the Causal Approach to Measurement Error," and "A Causal Approach to Nonrandom Measurement Errors"; Achen, "Mass Political Attitudes and the Survey Response"; Green, "On the Dimensionality of Public Sentiment toward Partisan and Ideological Groups"; Green and Citrin, "Measurement Error and the Structure of Attitudes."

37. Hayduk, *Structural Equation Modeling with LISREL*.

38. Kachigan, *Multivariate Statistical Analysis*.

39. Kerlinger, *Behavioral Research*, 180.

40. Responses to each of these questions are problematic because of the use of ordinal categories. Although Jöreskog and Sörbom, in *LISREL 7*, recommend that a polychoric correlation matrix be used with weighted least squares (WLS) under these conditions, that approach holds no special advantage here for several

reasons. First, polychoric correlation matrices do not alleviate the normality assumptions required by LISREL (a computer program acronym that stands for *LInear Structural RELationships*). Rather, the procedure requires the assumption that the true underlying variables are distributed multivariate normally and that the observed data appear as non-normal only because of poor and arbitrary classification into categories. Whether that assumption is accurate for most environmental data is unclear. While the measures used by Gallup in its environmental battery are negatively skewed, this may be due to poor cutpoints but more likely is due to a skewed underlying distribution since Americans seem overwhelmingly proenvironmental. As Hayduk warns, in *Structural Equation Modeling*, "if the problematic skewness really does originate from a skewed or otherwise nonmultivariate population distribution, we could be doing more harm than good by 'rectifying' the problem" (p. 329).

Second, use of any correlation matrix as an input matrix in LISREL leads to a loss of information about the real scales on which the indicators are based, interfering with the use of goodness-of-fit statistics and test statistics, such as chi-square. Because of this loss of information along the diagonal of the input matrix, use of a polychoric correlation matrix can make identification problems more severe and model convergence difficult.

Finally, a comparison of Pearson's and polychoric correlations in this case shows little difference in relative rank. True, Pearson's correlations are generally attenuated in comparison to polychoric correlations, but if we are interested in the rank order of coefficients (which measures correlate more highly than other coefficients), this information remains essentially the same. Consequently, a bivariate OLS regression of Pearson's correlations on corresponding polychoric correlations for the six measures of environmental concern used here yield a slope estimate of .93 and a R-square of .97. For all of these reasons, the use of maximum likelihood estimation is still an appropriate strategy.

41. As a preliminary step toward model specification, principal factor analysis was used on the full set of Gallup degree-of-concern items. Using a scree plot, eigenvalues dropped sharply after extracting the first factor and evened out in consecutive factors, with the first factor explaining 55 percent of the total variance in all variables. While high factor loadings for these items looked promising as evidence of near-perfect unidimensionality ranging in value from .63 to .78, those results are at best inconclusive and at worst misleading. Given that all thirteen questions were asked in close proximity using an identical response format, it is possible that this common response format drives the high factor loadings observed. To purge estimates of any systematic response bias, confirmatory factor analysis must be used instead.

42. The terms *skewness* and *kurtosis* are used by statisticians to characterize a variable's underlying distribution. Skewness represents the asymmetry of a distribution—that is, whether it leans to the right or to the left. Kurtosis (which derives from the Greek word for "bulginess") notes the size of a distribution's tails, which can cause a flat or thin appearance. While it is possible to employ an alternative estimator (as weighted least squares) if one or both of these

coefficients suggest non-normality, those techniques are more demanding computationally and are not clearly superior in performance. See Bollen, *Structural Equations with Latent Variables*; also Hayduk, *Structural Equation Modeling*.

43. Dividing chi-square by the degrees of freedom used is sometimes suggested as a less biased fit statistic than chi-square itself because of the latter's sensitivity to sample size. To suggest a good fit of the model to the data, the ratio calculated should be small, with values below three considered satisfactory. In this case, the multidimensional model's chi-square/degree-of-freedom ratio is 2.79. See Bollen and Long, *Testing Structural Equation Models*.

44. The closer the goodness-of-fit index (GFI) is to 1.00, the better the fit of the model to the data. The results in this case easily reach conventional levels of acceptance.

45. A nested chi-square difference test shows that the multidimensional model fails to provide a statistically significant improvement in fit over the unidimensional version. In this case, the unidimensional model is "nested" within the multidimensional model since it can be obtained by constraining one of the free parameters in the multidimensional model to be fixed—that is, it constrains the correlation between factors to be 1. With that in mind, the two models can be compared for goodness-of-fit using the following test statistic, where $\chi^2 = \chi_1^2 - \chi_2^2$, with $df_1 - df_2$ degrees of freedom:

	Chi-square	Degrees of freedom
Unidimensional model	21.10	8
Multidimensional model	−19.51	−7
	1.59	1

A chi-square of 1.59 with one degree of freedom is not statistically significant at conventional levels. In other words, relaxing the constraints of the unidimensional model does not result in a notable improvement in fit. See Jöreskog and Sörbom, *LISREL 7*.

46. While results are not reported in tabular form here, an identical model was used on a 1991 Gallup sample, and very similar results were obtained ($r = 0.91$). See Guber, "Environmental Concern and the Dimensionality Problem."

47. Given that only one measure of self-identification was available in the Gallup study (March 5–7, 2001), it was necessary to fix the error variance of this observed indicator to a predetermined value to achieve model identification. For this reason, each variable in the model was examined in conjunction with all remaining variables using a conventional method of reliability assessment (Cronbach's coefficient alpha), and the error variance for environmental self-identification in LISREL was set to a value that produced an equivalent reliability statistic.

48. In this case, several comparsions are in order. Summary correlations between additive scales created from each battery of questions ranged from just .36 to .43. Confirmatory factor analysis assuming no errors in measurement yielded corre-

lations within that same range. When random error was estimated, correlations improved to between .43 and .58 and finally to between .49 and .67 after controlling for nonrandom elements.

49. Buttel and Johnson, "Dimensions of Environmental Concern," 59.

50. deHaven-Smith, "Environmental Belief Systems," "Toward a Communicative Theory of Environmental Opinion," and *Environmental Concern in Florida and the Nation*.

51. Van Liere and Dunlap, "Environmental Concern."

52. Achen, "Mass Political Attitudes and the Survey Response"; Green, "On the Dimensionality of Public Sentiment toward Partisan and Ideological Groups"; Green and Citrin, "Measurement Error and the Structure of Attitudes."

53. It could be, for example, that truly multidimensional attitudes on the environment are *more* sophisticated, indicating that certain respondents are capable of drawing fine distinctions between different and complex environmental issues.

54. Pierce and Lovrich, "Belief Systems Concerning the Environment," 282.

55. Achen, "Mass Political Attitudes and the Survey Response," 1231. His criticism seems particularly valid here given evidence suggesting that most Americans are ill-informed on environmental issues. For example, the National Environmental Education and Training Foundation's (NEETF) annual "report card" on environmental knowledge finds that "Americans lack the basic knowledge and are unprepared to respond to the major environmental challenges we face in the twenty-first century" (available online at ⟨http://www.neetf.org/roper/roper.shtm⟩). Others, too argue that the "environmental risks that frighten people most rarely matter most." See "Green Choices, Hard Choices," 11; and Rauch, "There's Smoke in the Air, But It Isn't All Pollution," B1. For a more scholarly interpretation of the same basic point, refer to Slovic, "Perception of Risk."

Chapter 6

1. Carmines and Stimson, "The Two Faces of Issue Voting."

2. Mitchell, "Public Opinion and Environmental Politics in the 1970s and 1980s," in Vig and Kraft, eds., *Environmental Policy in the 1980s*, 55; Dunlap, "Polls, Pollution, and Politics Revisited" and "Public Opinion in the 1980s."

3. Taylor, "Campaign Trail Littered with Environmental Wrecks," 7b; Zaller, *Report on 1991 Pilot Items on Environment*.

4. Taylor, "Campaign Trail Littered," 7b.

5. In all fairness, it should be noted that both Mitchell, in "Public Opinion and Environmental Politics," and Dunlap, in "Polls, Polution, and Politics Revisited" and "Public Opinion in the 1980s," suggest possible reasons for the weak electoral impact of environmentalism, with Mitchell stressing high public support for environmental goals in conjunction with low issue salience and Dunlap emphasizing a "permissive consensus" that affords elected officials considerable

flexibility and independence in pursuing environmental policies. Because of a lack of available data, however, neither is able to examine issue voting on the environment directly.

6. Kriz, "The Green Card," 2262, and "Slinging Earth," 958; Garland, "Come Winter," 66; Borosage and Greenberg, "Why Did Clinton Win?"; St. Clair, "The Twilight of 'Gang Green,'" 14; Bedard, "Vote Green," 9.

7. Carmines and Stimson, "The Two Faces of Issue Voting," 79.

8. Downs, "Up and Down with Ecology"; Nie, Verba, and Petrocik, *The Changing American Voter*; Margolis, "From Confusion to Confusion."

9. Shabecoff, "Shades of Green in the Presidential Campaign," 73–74.

10. Dunlap and Allen, "Partisan Differences on Environmental Issues."

11. For a helpful overview of the literature, see Kamieniecki, "Political Parties and Environmental Policy," in Lester, ed., *Environmental Politics and Policy*; also Cooley and Wandesforde-Smith, eds. *Congress and the Environment*; Dunlap and Gale, "Party Membership and Environmental Politics"; Dunlap and Allen, "Partisan Differences"; Calvert, "The Social and Ideological Bases of Support for Environmental Legislation."

12. Charles R. Shipan, and William R. Lowry, "Environmental Policy and Party Divergence in Congress," 245–263.

13. Lake, "The Environmental Mandate," 230–231.

14. Zaller, *Report on 1991 Pilot Items on Environment*.

15. Kriz, "The Green Card" and "Slinging Earth."

16. Schuman and Presser, *Questions and Answers in Attitude Surveys*.

17. Rabinowitz, Prothro, and Jacoby, "Salience as a Factor in the Impact of Issues on Candidate Elections."

18. While there is some disagreement in the literature regarding the cognitive process of issue voting ("spatial" or "directional"), both models require that voters perceive differences between the candidates on issues that matter to them. See, for example, Downs, *An Economic Theory of Democracy*; Enelow and Hinich, *The Spatial Theory of Voting*; Rabinowitz and Macdonald, "A Directional Theory of Issue Voting"; Macdonald, Rabinowitz, and Listhaug, "Political Sophistication and Models of Issue Voting" and "On Attempting to Rehabilitate the Proximity Model"; Westholm, "Distance versus Direction."

19. Campbell, Converse, Miller and Stokes, *The American Voter*.

20. Campbell et al., *The American Voter*, 179; see also Carmines and Stimson, "The Two Faces of Issue Voting," 82.

21. Brody and Page, "Comment"; Pomper, "From Confusion to Clarity" and *Voters' Choice*; also Nie, Verba, and Petrocik, *The Changing American Voter*.

22. In this sense, the environment may fail to influence individual vote choice because it is a "valence" issue—that is, one on which nearly everyone agrees,

either in practice or in rhetoric. See Stokes, "Spatial Models of Party Competition." Asked once about his thoughts on Earth Day, former House Speaker Newt Gingrich (Republican—Georgia) quipped it was the day when "Republicans dress up in drag and pretend that they're environmentalists" (Kriz, "Slinging Earth," 1957).

23. Peffley, Feldman, and Sigelman, "Economic Conditions and Party Competence."

24. Berinsky and Rosenstone, "Evaluation of Environmental Policy Items on the 1995 NES Pilot Study."

25. Mastio, "The GOP's Enviro-Rut," 19.

26. The choice here of a candidate evaluation scale over a simple vote preference is consistent with recent trends in the field and as Rabinowitz et al. ("Salience as a Factor," 45) note, allows "a range of response rather than being restricted to for or against." Zaller's model of environmental preferences in a 1991 NES Pilot Study report takes the same approach; see Zaller, *Report on 1991 Pilot Items on Environment.*

27. For example, Campbell et al., *The American Voter*; also Aldrich, Sullivan, and Borgida, "Foreign Affairs and Issue Voting."

28. Carmines and Stimson, "The Two Faces of Issue Voting," 82.

29. Why do voters have difficulty in distinguishing between the policy positions of the candidates? Perhaps it is because they are not well informed on environmental issues. After all, many of those polled in the National Election Study (NES) were "not very certain" of their placement of Bill Clinton (30 percent) or Bob Dole (40 percent) on the environment and economy scale. Those respondents were, on average, more likely to position the candidates toward the center of the continuum within close proximity of one another. It is interesting to note, however, that the degree of certainty respondents feel about the candidates' issue positions is not strongly related to any factors that seem instinctive or obvious—not to partisanship or education, or to the amount of attention paid to the presidential campaign. As a result, this problem may not be easy for candidates to rectify.

30. Since this analysis closely follows the work of Aldrich et al., further detail on the logic of these "cumulative conditions" can be found in "Foreign Affairs."

31. Smith, "Abortion Attitudes and Vote Choice in the 1984 and 1988 Presidential Elections."

32. Unfortunately, it is not possible to differentiate between "true" Independents and more partisan "leaners" in this model. Small sample sizes simply do not permit a more refined analysis.

33. Harris and Nakashima, "Gore's Greenness Fades," A1.

34. Callahan, "Environment Is 'Sleeper' Issue of 2000 Campaign."

35. Because of a series of experiments involving question wording and survey administration, a fully specified regression model (like that reported in table 6.3) is not possible using 2000 NES data. The sample sizes for that model, using a

listwise deletion of missing values, are simply too small to support statistical analysis.

36. Nieves, "Conversation/Ralph Nader," D7.

37. "Mr. Nader's Electoral Mischief," A34.

38. McFeatters, "Green Party Candidate Continues to Criticize Bush and, Particularly, Gore," A8. The 2.8 percent of self-reported votes that Nader received among NES respondents in 2000 was a near perfect reflection of his national vote total. Unfortunately, in public opinion polls small numbers translate into small sample sizes. In this case, just thirty-three of those polled reported they voted for Nader—a subset far too small to support reliable statistical analysis.

39. Official election returns provided by the Federal Election Commission (FEC) indicate that Nader received 97,488 votes in Florida—a state in which George W. Bush's official margin of victory was just 537.

40. "Mr. Nader's Misguided Crusade," A24.

41. Dunlap and Scarce, "The Polls."

42. Dunlap, "Public Opinion and Environmental Policy," in Lester, ed., *Environmental Politics and Policy*, 130.

43. Bosso, "After the Movement," in Vig and Kraft, eds. *Environmental Policy in the 1990s*; also Kriz, "The Green Card."

44. Purdy, "Planet Bush, Planet Gore," 34.

45. Purdy, "Planet Bush, Planet Gore," 34.

46. Purdy, "Planet Bush, Planet Gore," 34.

47. Schneider, "Everybody's an Environmentalist Now."

48. Bragdon and Donovan, "Voter's Concerns Are Turning the Political Agenda Green," 186.

49. Ridgeway, "It Isn't Easy Voting Green," 144–145.

50. Mitchell, "Public Opinion and the Green Lobby," in Vig and Kraft, eds., *Environmental Policy in the 1990s*; Mitchell, Mertig, and Dunlap, "Twenty Years of Environmental Mobilization," in Dunlap and Mertiq, eds., *American Environmentalism*.

51. Paige, "The 'Greening' of Government," 16.

52. McCloskey, "A Second-Order Issue," 2.

53. Lake, "The Environmental Mandate"; also Kahn and Matsusaka, "Demand for Environmental Goods."

Chapter 7

1. Dunlap, "Polls, Pollution, and Politics Revisited," 13.

2. Zisk, *Money, Media, and the Grassroots*, 161.

3. Magleby, *Direct Legislation*.

4. Lake, "The Environmental Mandate"; Johnson, "Citizens Initiate Ballot Measures."

5. Taylor, "Campaign Trail Littered with Environmental Wrecks," 7b.

6. Ridenour, "The Mouse that Squeaked."

7. Lake, "The Environmental Mandate," 222.

8. Magleby, *Direct Legislation*, 168.

9. Lee, "California," in Butler and Ranney, eds., *Referendums*.

10. Bone and Benedict, "Perspectives on Direct Legislation," 47.

11. Zisk, *Money, Media, and the Grassroots*.

12. Magleby, *Direct Legislation*.

13. Lau, "Two Explanations for Negativity Effects in Political Behavior."

14. Magleby, in *Direct Legislation*, finds, for instance, that while just 26 percent of all citizens shift their vote intention in candidate elections, 70 percent of voters change their opinions on ballot measures, most commonly from "soft" early support to outright opposition by election day.

15. Mueller, "Voting on the Propositions," 1211.

16. Lowenstein, "Campaign Spending and Ballot Propositions"; also Bone and Benedict, "Perspectives on Direct Legislation"; Magleby, *Direct Legislation*.

17. Magleby, "Direct Legislation in the American States," in Butler and Ranney, eds. *Referendums around the World*, 251.

18. See Zisk, *Money, Media, and the Grassroots*, 165.

19. Lee, "California," in Butler and Ranney, eds., *Referendums*.

20. Magleby, "Opinion Formation and Opinion Change in Ballot Proposition Campaigns," in Margolis and Mauser, eds., *Manipulating Public Opinion*, 113; also Bowler, Dovovan, and Happ, "Ballot Propositions and Information Costs"; Darcy and McAllister, "Ballot Position Effects."

21. Magleby, *Direct Legislation*, 54.

22. Baus and Ross, *Politics Battle Plan*.

23. Zisk, *Money, Media, and the Grassroots*.

24. Bone and Benedict, "Perspectives on Direct Legislation," 346.

25. Bowler and Donovan, "Economic Conditions and Voting on Ballot Propositions."

26. Rourke, Hiskes, and Zirakzadeh, *Direct Democracy and International Politics*, 42–43.

27. A database on statewide initiatives from the Initiative and Referendum Institute is available online at ⟨http://www.iandrinstitute.org⟩.

28. Graham, *A Compilation of Statewide Initiative Proposals Appearing on Ballots through 1976*; Ranney, "The United States of America," in Butler and Ranney, eds., *Referendums*, 80.

29. For example, Magleby, *Direct Legislation*; Zisk, *Money, Media, and the Grassroots*; Bowler and Donovan, "Economic Conditions and Voting on Ballot Propositions."

30. Hawaii, South Carolina, Tennessee, and Virginia were unable to provide data on ballot propositions.

31. Also excluded from this analysis were non-environmental measures nonetheless supported by environmental groups, including (for example) campaign finance reform or regulations regarding the initiative process and the number of signatures required to certify a petition.

32. Magleby, "Direct Legislation in the American States," in Butler and Ranney, eds., *Referendums around the World*, 251.

33. It is also possible that the threat of impending war with Iraq during Desert Shield and Desert Storm contributed to voter pessimism in 1990.

34. A *dummy variable* is a dichotomous variable—that is, one that can take on only two possible values, 0 and 1. See Berry and Sanders, *Understanding Multivariate Research*.

35. The estimate that environmental ballot propositions represent roughly 10 percent of the total is based on the author's own data collected from state election offices. It is also consistent with complete tallies from the four states examined in this chapter. In Colorado, Massachusetts, Oregon, and South Dakota combined, 549 ballot questions were offered to voters between 1964 and 2000. Forty-nine of those—or 8.9 percent—were on environmental subjects.

36. Zisk, *Money, Media, and the Grassroots*.

37. Saffire, "Has the Time Come for Deposit Legislation?," 25.

38. Zisk, *Money, Media, and the Grassroots*; Johnson, "Citizens Initiate Ballot Measures."

39. Russell, "Warnings, Warnings Everywhere," Z12, Z14.

40. Lovett, "Proposition 65 Comes of Age," 26.

41. Russell, "Warnings, Warnings Everywhere," Z12.

42. Epstein, "Campaign to Defeat Proposition 65," 6 (emphasis added); also LaVally, "Proposition 65 Opponents Outspent Successful Foes."

43. Locke, "California Oil Industry Joins Others in Wariness over Proposition 65," 8.

44. Locke, "California Oil Industry Joins Others," 8.

45. Jacobs, "Prop. 65," 1.

46. Skelton, "Deukmejian Opposes 3 Controversial Propositions," 3.

47. Field poll data are archived at the University of California and at the Roper Center for Public Opinion Research at the University of Connecticut. All three polls (USCA 86-04, USCA 86-05, USCA 86-06) were administered to adult residents of California selected by random-digit dialing. The first (conducted July 24

to August 4, 1986) had a sample size of 1,028; the second (September 24 to October 2, 1986) had a sample size of 1,023; and the last poll (October 29 to 30, 1986) had a sample size of 701. For more information on the Field Institute, see ⟨http://www.field.com/fieldpoll⟩.

48. Locke, "Success of Proposition 65," 8.

49. Epstein, "Campaign to Defeat Proposition 65."

50. Zisk, *Money, Media, and the Grassroots.*

51. Nicholl, "Son of David and Goliath," 17 (emphasis added).

52. Allswang, *California Initiatives and Referendums*; also Nicholl, "Son of David and Goliath."

53. Rauber, "Losing the Initiative?," 20.

54. Lacayo, "Green Ballots vs. Greenbacks," 44; also Rauber, "Losing the Initiative?"

55. Casuso, "Black Tuesday for Big Green Backers," 6.

56. Allen, "High Supply, Low Demand Hurt Effort to Recycle," 27.

57. Dumanoski, "Environment Not Gaining Ground during Campaign," 1 and "Advertising Blitz Erodes Support for Question 3," 17.

58. Allen, "High Supply"; Leaversuch, "Massachusetts Votes Down Drastic Measure to Reduce Package Waste," 37.

59. Allen, "High Supply"; Dumanoski, "Environment Not Gaining Ground during Campaign" and "Advertising Blitz Erodes Support for Question 3."

60. Magleby, *Direct Legislation.*

61. Designed for the Massachusetts Public Interest Research Group (MassPIRG), the environmental sponsors of Question 3, the Marttila & Kiley survey (#MK 92110) was administered by telephone from February 11 to 13, 1992. A sample of 402 Massachusetts voters was generated using a random-probability method that included unlisted telephone numbers. The sample was stratified according to county, and gender quotas were observed. The data used here were provided to the author courtesy of Marttila Communications Group, Inc. (1 Beacon Street, Boston, MA 02111). See also ⟨http://www.masspirg.org⟩.

62. Magleby, *Direct Legislation.*

63. Leaversuch, "Massachusetts Votes Down Drastic Measure to Reduce Package Waste."

64. Gilmore, "Nation's Voters Decide to Keep their Wallets Green," A5; Mathews, "The Big, the Green, the Political," A19.

65. Mathews, "The Big, the Green, the Political," A19.

66. Taylor, "Campaign Trail Littered," 7b.

67. Magleby, *Direct Legislation*, 144.

68. Carmines and Stimson, "The Two Faces of Issue Voting."

69. Zisk, *Money, Media, and the Grassroots.*

Chapter 8

1. Carson and Moulden, *Green Is Gold*, 4.

2. The supply of "environmentally friendly" products appears to be a relatively recent phenomenon. While fewer than 2 percent of all new products introduced in 1986 boasted environmental claims, "green" products accounted for more than 10 percent by 1994. See Winski, "Big Prizes, But No Easy Answers," GR-3; Gray-Lee, Scammon, and Mayer, "Review of Legal Standards for Environmental Marketing Claims"; Scammon and Mayer, "Agency Review of Environmental Marketing Claims."

3. Ottman, "Environmentalism Will Be the Trend of the '90s," 13. The most famous of these partnerships between a national environmental group and a major corporation occurred in 1990 between the Environmental Defense Fund and the McDonald's Corporation. The collaboration produced a Waste Reduction Action Plan that outlined forty-two ways in which the popular restaurant chain could reduce its generation of solid waste, including the substitution of quilted paper wraps for polystyrene clamshell containers. By 1995 the plan had grown to nearly a hundred initiatives, with McDonald's using its purchasing power to motivate suppliers to meet those standards, resulting in both corporate cost savings and reduced waste. See Holusha, "Talking Deals," D2.

4. Mitchell, "Public Opinion and Environmental Politics in the 1970s and 1980s," in Vig and Kraft, eds., *Environmental Policy in the 1980s*; Dunlap, "Polls, Pollution, and Politics Revisited" and "Public Opinion in the 1980s."

5. Buchanan, "Individual Choice in Voting and the Market"; Buchanan and Tullock, *The Calculus of Consent*; Wilson and Banfield, "Public Regardingness as a Value Premise in Voting Behavior" and "Voting Behavior on Municipal Public Expenditures," in Margolis, ed., *The Public Economy of Urban Communities*; Martinez-Vazquez, "Selfishness versus Public 'Regardingness' in Voting Behavior."

6. Sunstein, "Remaking Regulation," 74.

7. Downs, *An Economic Theory of Democracy*; Deacon and Shapiro, "Private Preference for Collective Goods Revealed through Voting on Referenda."

8. Green, "The Price Elasticity of Mass Preferences," 128.

9. Sears and Huddy, "Bilingual Education."

10. Kinder and Sanders, "Mimicking Political Debate With Survey Questions."

11. Kluegel and Smith, "Whites' Beliefs about Blacks' Opportunity."

12. Beck and Dye, "Sources of Public Opinion on Taxes."

13. See, Wilson and Banfield, "Public Regardingness" and "Voting Behavior on Municipal Public Expenditures," in Margolis, ed., *The Public Economy of Urban Communities*; also Kelman, *What Price Incentives?*; Etzioni, *The Moral Dimension*; Monroe, "John Donne's People." Although Wilson and Banfield did not directly link their research to public choice theory, their conceptualization of

"public-regarding" versus "private-regarding" behavior parallels the latter's definition of self-interest and altruism. See Wilson and Banfield, "Public Regardingness"; also Brodsky and Thompson, "Ethos, Public Choice, and Referendum Voting," 289.

14. Buchanan, "Individual Choice," 336.

15. Sunstein, "Remaking Regulation," 75.

16. Gerlak and Natali, *Taking the Initiative*, II.

17. A comprehensive guide to green products can be found online at ⟨http://www.buygreen.com⟩.

18. Schwartz and Miller, "The Earth's Best Friends"; Coddington, *Environmental Marketing*; Stisser, "A Deeper Shade of Green."

19. Morris, Hastak, and Mazis, "Consumer Comprehension of Environmental Advertising and Labeling Claims"; Wasik, "Market Is Confusing, But Patience Will Pay Off," 16.

20. Coddington, *Environmental Marketing*.

21. Some scholars argue, for example, that altruistic behavior is frequently motivated by economic behavior. See Holmes, "Self-Interest, Altruism, and Health-Risk Reduction."

22. Oskamp, Harrington, Edwards, Sherwood, Okuda, and Swanson, "Factors Influencing Household Recycling Behavior."

23. Milbrath, *Environmentalists*; also Dunlap, "Public Opinion in the 1980s."

24. De Young, "Encouraging Environmentally Appropriate Behavior" and "Some Psychological Aspects of Recycling."

25. Geller, Winett and Everett, *Preserving the Environment*.

26. Hopper and Nielsen, "Recycling as Altruistic Behavior."

27. Black, "Attitudinal, Normative, and Economic Factors."

28. Baldassare and Katz, "The Personal Threat of Environmental Problems as a Predictor of Environmental Practices."

29. Dunlap and Beus, "Understanding Public Concerns about Pesticides."

30. Rohrschneider, "Citizens' Attitudes toward Environmental Issues."

31. Constantini and Hanf, "Environmental Concern and Lake Tahoe"; Dillman and Christenson, "The Public Value for Pollution Control," in Burch et al., eds., *Social Behavior, Natural Resources, and the Environment*; Tognacci et al., "Environmental Quality"; Dunlap, "The Impact of Political Orientation on Environmental Attitudes and Action"; Buttel and Flinn, "The Politics of Environmental Concern"; Van Liere and Dunlap, "The Social Bases of Environmental Concern"; Samdahl and Robertson, "Social Determinants of Environmental Concern."

32. Buchanan, "Individual Choice."

33. That result depends, however, on the truthfulness and accuracy of the environmental claims made by manufacturers. The mass marketing of environmental values may have popularized terms such as *environmentally friendly*, *biodegradable*, and *ozone safe*, but some Americans remain skeptical of the merits of those claims. See Carlson, Grove, and Kangun, "A Content Analysis of Environmental Advertising Claims"; see also Schrum, Lowrey, and McCarty, "Recycling as a Marketing Problem."

34. Schwartz and Miller, "The Earth's Best Friends"; Coddington, *Environmental Marketing*.

35. Kinnear, Taylor, and Ahmed, "Ecologically Concerned Consumers"; Schwepker and Cornwell, "An Examination of Ecologically Concerned Consumers"; see also Schrum, Lowrey, and McCarty, "Recycling as a Marketing Problem."

36. Popular discount retailers Kmart, Wal-Mart, and Target all have used green shelf tags to varying degrees in promoting environmentally safe products. See Halverson, "Big Three Take High Road on Environmental Front," 72.

37. Schuhwerk and Lefkoff-Hagius, "Green or Non-Green?"

38. See Singhapakdi and LaTour, "The Link between Social Responsibility Orientation, Motive Appeals, and Voting Intention."

39. According to Donald Green and Irene Blair, "instrumental" appeals are those that call attention to the direct benefits of a particular good or policy. In contrast, "symbolic" appeals draw attention to the gratification an individual may receive by participating in or expressing support for a given cause or product. See Green and Blair, "Framing and the Price Elasticity of Private and Public Goods."

40. Dunlap and Scarce, "The Polls."

41. One such exception is the purchase of a less "gas guzzling" automobile—it ranked eighth out of eleven activities in a 1991 Gallup study (April 11–14). Industry polling suggests a straightforward explanation. With reasonable gasoline prices, fuel efficiency is simply not a top priority for most consumers. Safety, durability, seating capacity, and style are all factors that in recent years have pushed Americans away from small, lightweight cars toward heavy sport-utility vehicles, minivans, and pickup trucks (The Gallup Organization, April 11–14, 1991 [$n = 1007$].

42. See Green and Blair, "Framing and the Price Elasticity of Private and Public Goods."

43. See Rohrschneider, "Citizens' Attitudes toward Environmental Issues," 364.

44. Kinder and Kiewiet, "Sociotropic Politics."

45. Sears, Lau, Tyler, and Allen, "Self-Interest vs. Symbolic Politics."

46. *Multicollinearity* is a term used to represent the presence of a relationship between different independent variables. See Berry and Sanders, *Understanding Multivariate Research*, 43–44.

47. See, for example, Huber, "How to Love Butterflies and be a Conservative."

48. Buchanan and Tullock, *The Calculus of Consent*, 37.

49. Buchanan and Tullock, *The Calculus of Consent*, 37.

50. Kinnear, Taylor, and Ahmed, "Ecologically Concerned Consumers," 24; see also Ellen, Wiener, and Cobb-Walgren, "The Role of Perceived Consumer Effectiveness"; Berger and Corbin, "Perceived Consumer Effectiveness and Faith in Others."

51. Green, "The Price Elasticity of Mass Preferences," 139.

52. Schuhwerk and Lefkoff-Hagius, "Green or Non-Green?"

53. Buchanan and Tullock, *The Calculus of Consent*, 21.

54. Citrin, Reingold, and Green, "American Identity and the Politics of Ethnic Change"; Citrin, Reingold, Walters, and Green, "The 'Official English' Movement."

55. Buchanan and Tullock, *The Calculus of Consent*, 20.

Conclusion

1. Shabecoff, *Earth Rising*, 26.

2. Shabecoff, "Shades of Green in the Presidential Campaign," 73.

3. A strong majority of those polled (83 percent) in an April 2000 poll agreed with the goals of the environmental movement either "strongly" or "somewhat," ranking it just behind the civil rights (86 percent) and women's rights (85 percent) movements. Respondents were somewhat less sure of its success, however, with just 30 percent believing it had accomplished "a great deal," compared once again to the civil rights movement (50 percent) and the women's movement (42 percent). See Dunlap, "Americans Have Positive Image of the Environmental Movement"; see also Mertig and Dunlap, "Public Approval of Environmental Protection and Other New Social Movement Goals"; Dowie, "American Environmentalism" and *Losing Ground*; and Rubin, *The Green Crusade*.

4. The phrase "green revolution" in the context of U.S. environmental politics is inspired by Kirkpatrick Sale's 1993 book of the same name.

5. Wiley, "Coming to Terms," 28.

6. Purdy, "Shades of Green," 11. See also Schroeder, "Clear Consensus, Ambiguous Commitment," 1876.

7. Wiley, "Coming to Terms," 28.

8. The term *sympathetic public* is used in Morrison, "How and Why Environmental Consciousness Has Trickled Down," in Schnaiberg, Watts, and Zimmermann, eds., *Distributional Conflict in Environmental-Resource Policy*; see also Dunlap, "Public Opinion and Environmental Policy," in Lester, ed., *Environmental Politics and Policy*.

9. Dunlap and Mertig, "The Evolution of the U.S. Environmental Movement from 1970 to 1990."

10. The strength of the relationship between environmental attitudes and the economy creates an discomforting irony for environmental ideologues. It implies that public support for environmental reform is contingent (at least in part) on the very economic growth and prosperity some fear caused those conditions from the start. For a classic argument on this topic, refer to Meadows et al., *The Limits to Growth*.

11. For example, the environmental risks that frighten people are rarely the ones that matter most. See, for example, "Green Choices, Hard Choices." For more detailed scholarship, refer to Slovic, "Perception of Risk."

12. Kempton, Boster, and Hartley, *Environmental Values in American Culture*, 220.

13. Kempton, Boster, and Hartley, *Environmental Values in American Culture*, 220; also Wall, "Barriers to Individual Environmental Action."

14. Teles, "Think Local, Act Local," 28.

15. Schuman and Presser, *Questions and Answers in Attitude Surveys*; see also Rabinowitz, Prothro, and Jacoby, "Salience as a Factor in the Impact of Issues on Candidate Elections."

16. The news media might be at least partially responsible for perpetuating environmental myths and misconceptions, but there is spirited disagreement about whether reporters over- or under-emphasize the seriousness of those problems. For contrasting viewpoints, refer to Michaels, "Reports Cry Wolf about Environment"; and Ruben, "Back Talk."

17. The examples cited above are drawn from the National Environmental Education and Training Foundation's (NEETF) 1998 *National Report Card on Environmental Attitudes, Knowledge, and Behavior*. For more information, including an online version of this report, visit the National Environmental Education and Training Foundation's Web site at ⟨http://www.neetf.org⟩.

18. Evidence on this point is mixed. While V. Kerry Smith finds evidence that education positively influences environmental action, Riley E. Dunlap is "skeptical that public support for environmental protection will hinge upon lay-persons becoming highly informed" about technical issues like global warming. Matthias Finger, too, argues that knowledge and information "predict little of the variability in most forms of environmental behavior." See Smith, "Does Education Induce People to Improve the Environment?"; Finger, "From Knowledge to Action?"; Dunlap, "Lay Perceptions of Global Risk."

19. See, in particular, Olson, *The Logic of Collective Action*.

20. Ridley and Low, "Can Selfishness Save the Environment?"

21. Hardin, "The Tragedy of the Commons."

22. Wall, "Barriers to Individual Environmental Action," 483; see also Sunstein, "Remaking Regulation."

23. Goodrich, "Just a Few Words of Optimism from Green Enemy No. 1," E6.

24. Knickerbocker, "Is It Wrong to be Optimistic about the Environment?," 14.

25. Mathews, "Outlook Not So Gloomy," B9.

26. Rauch, "There's Smog in the Air, But It Isn't All Pollution," B01. Mark Shields argues that "the reason no one knows what progress we've made is that there is a conspiracy between Republicans, who don't want to admit that government works, and Democrats, who don't want to acknowledge that most of the progress occurred under Republican presidents. So they both keep quiet about it." Quoted in "Environmental Protection: Is the Public Willing to Pay?," 15.

27. Easterbrook, *A Moment on the Earth*, xviii.

28. Easterbrook, *A Moment on the Earth*, xviii. See also Lomborg, *The Skeptical Environmentalist*.

29. Immerwahr, *Waiting for a Signal*.

30. For example, critics have long argued that environmental education programs in schools serve as a form of political indoctrination, frightening children with inaccurate tales of environmental disaster. For a balanced discussion, see Hungerford and Lewis, "Is Environmental Education Scaring Our Children to Death?"; also Sanera and Shaw, *Facts, Not Fear*.

31. Immerwahr, *Waiting for a Signal*.

32. Ridley and Low, "Can Selfishness Save the Environment?," 77.

33. Quoted in Braile, "Book Review: The Land that Could Be," Books-3. For far more detail on the possibilities of "civic environmentalism," see Shutkin, *The Land That Could Be*; also Knopman, Susman, and Landy, "Civic Environmentalism."

34. Shutkin, *The Land That Could Be*, 22.

35. Wilson, "Why They Don't Campaign about the Environment." In this sense, environmental protection has become a valence issue, one "where virtually everyone supports the goal, thus confining potential disagreement to the means by which these ends can be achieved." See Hibbing and Theiss-Morse, *Congress as Public Enemy*, 55; also Stokes, "Spatial Models of Party Competition."

36. See, for example, Rauch, "There's Smog in the Air," B01.

37. These figures are drawn from a *Newsweek* poll conducted by Princeton Survey Research Association, April 13 and 14, 2000, with a sample size of 752 adults nationwide and a margin of error of plus or minus four percentage points. The wording of the question was this: "Since the first Earth Day was held thirty years ago, how much progress—if any—do you think has been made toward solving environmental problems?" Tabulated responses were 18 percent "major progress," 52 percent "minor progress," 7 percent "no progress," 16 percent "gotten worse," and 7 percent "don't know." A second study conducted at the same time by Environmental Defense found even greater pessimism, with a majority believing that conditions had grown worse. Those results (available in

a press release online at ⟨http://environmentaldefense.org⟩) are also discussed in Kates, "Has the Environment Improved?"

38. Norman Levitt, as quoted in Malik, "Why Science Needs Protection," 13; for Levitt's full argument, see Levitt, *Prometheus Bedeviled*.

39. Spangler, *Methods, Problems and Issues in Environmental Cost-Benefit Analysis of Nuclear Power Plant Alternatives*, 3.

40. Kasper, "Perceptions of Risk and Their Effects on Decision Making," and Slovic, Fischhoff, and Lichtenstein, "Facts and Fears," both in Schwing and Albers, eds., *Societal Risk Assessment*.

41. Breyer, *Breaking the Vicious Circle*. But is Breyer right about the science of risk regulation? Are average Americans simply ill-informed, irrational, or inconsistent in the risks they perceive? Perhaps not. Studies show that people are reasonably accurate when asked to estimate fatality rates. The same group that awards nuclear power low yearly fatality estimates, however, still asserts nuclear power to be among the riskiest technologies of modern life. In short, while experts tend to equate risk with fatality estimates, lay people view risk as something more than just fatality, exposure, or reduced life expectancy. They depend heavily on inferences or heuristics that allow them to connect past experiences or available knowledge to the situation at hand, often evaluating risk based on more complex criteria, such as involuntary exposure, immediacy, dread consequences, and lack of personal control over harm. See Slovic, Fischhoff, and Lichtenstein, "Facts and Fears," in Schwing and Albers, eds., *Societal Risk Assessment*, and Slovic, "Perception of Risk."

42. Teles, "Think Local, Act Local," 28.

43. Carson and Moulden, *Green Is Gold*.

44. Schultze, *The Public Use of Private Interest*, 18.

45. Anderson, "Market Ecology Approach Hailed," C7; Eddy, "Loveland Sets Bar High for Recycling," A17.

46. Kelman, *What Price Incentives?*

47. Ellis and Thompson, "Culture and the Environment in the Pacific Northwest," 892.

48. Herndl and Brown, *Green Culture*.

49. Citrin, Reingold, and Green, "American Identity and the Politics of Ethnic Change."

50. Magleby, *Direct Legislation*.

51. Key, *Public Opinion and American Democracy*, 27.

52. After all, as Jedediah Purdy notes, most environmental laws date from the earliest days of the movement in the 1970s, and no major piece of environmental legislation has been passed since 1990. To make matters worse, he says, nearly "everyone concedes that the old laws are imperfect, often poorly enforced, and sometimes downright perverse." See Purdy, "Planet Bush, Planet Gore," 34.

53. For example, as David Broder notes: "At one level, the environmentalists have swept away all opposition. The 'conservation ethic' has become one of the fixed guiding stars of American politics—a 'value question' that permits only one answer from anyone who hopes to be part of the public dialogue.... The argument is no longer about values. That's over, and the environmentalists have won. The argument is now about policies. And those with the best evidence and the best arguments, not just the purest hearts, will prevail." See Broder, "Beyond Folk Songs and Flowers," B7; see also Schroeder, "Clear Consensus," 1879.

54. Dowie, "American Environmentalism," 70.

55. Gup, "Owl vs. Man," 62.

56. See Lake, "The Environmental Mandate," 230–231; also Cooley and Wandesforde-Smith, eds., *Congress and the Environment*; Dunlap and Gale, "Party Membership and Environmental Politics"; Dunlap and Allen, "Partisan Differences on Environmental Issues"; Calvert, "The Social and Ideological Bases of Support for Environmental Legislation."

57. See also Campbell, Converse, Miller, and Stokes, *The American Voter*. Richard Fenno writes on a similar theme. He argues that "voting leeway" occurs when constituents given a favorable evaluation of an incumbent's performance regardless of their voting record. Fenno, *Home Style*.

58. Purdy, "Planet Bush, Planet Gore," 34.

59. Hershkowitz, "Green vs. Greenbacks," GR10.

References

Achen, Christopher. 1975. Mass Political Attitudes and the Survey Response. *American Political Science Review*, 69: 1218–1231.

Albrecht, Don, Gordon Bultena, Eric Hoiberg, and Peter Nowak. 1982. The New Environmental Paradigm Scale. *Journal of Environmental Education*, 13: 39–43.

Albrecht, Stan L., and Armond L. Mauss. 1975. The Environment as a Social Problem. In Armond L. Mauss, ed., *Social Problems as Social Movements* (556–605). Philadelphia: Lippincott.

Aldrich, John H., and Forrest D. Nelson. 1984. *Linear Probability, Logit, and Probit Models*. Beverly Hills: Sage.

Aldrich, John H., John L. Sullivan, and Eugene Borgida. 1989. Foreign Affairs and Issue Voting: Do Presidential Candidates "Waltz before a Blind Audience?" *American Political Science Review*, 83(1): 123–141.

Alexander, Charles P. 1992. Gunning for the Greens. *Time* (February 3): 50.

Allan, Stuart, Barbara Adam, and Cynthia Carter, eds. 2000. *Environmental Risks and the Media*. New York: Routledge.

Allen, Scott. 1992. High Supply, Low Demand Hurt Effort to Recycle. *Boston Globe*, August 30, 27.

Allswang, John M. 1991. *California Initiatives and Referendums, 1912–1990: A Survey and Guide to the Research*. Los Angeles: Edmund G. "Pat" Brown Institute of Public Affairs, California State University.

Alpert, Bruce. 1995. Voters Support Rules on Pollution; GOP Attacks Not Popular. *Times-Picayune*, July 24, A1.

Anderson, Robert. 1991. Market Ecology Approach Hailed: Conservatives Support Incentives, Not Rules. *Seattle Times*, August 30, C7.

Anthony, Carl. 1999. Race, Justice, and Sprawl. *Forum for Applied Research and Public Policy*, 24: 97.

Ayres, Ed. 1999. The Worldwatch Report: Why Are We Not Astonished? *ENN News* (May 12). 〈http://www.enn.com/enn-features-archive/1999/05/051299/worldwatch_3141.asp〉.

Baldassare, Mark, and Cheryl Katz. 1992. The Personal Threat of Environmental Problems as Predictor of Environmental Practices. *Environment and Behavior*, 24: 602–616.

Baus, H. M., and W. B. Ross. 1968. *Politics Battle Plan*. New York: Macmillan Press.

Beck, Paul Allen, and Thomas R. Dye. 1982. Sources of Public Opinion on Taxes: The Florida Case. *Journal of Politics*, 44: 172–182.

Bedard, Paul. 1998. Vote Green: Democratic Tactics to Win Voters. *U.S. News & World Report* (October 19): 9.

Begley, Sharon, Mary Hager, and Lynda Wright. 1990. The Selling of Earth Day. *Newsweek* (March 26): 60.

Belden and Russonello. 1996. The Ecology Poll (USBELDEN96-ECOLOGY) [datafile]. February 29–March 12, $n = 2,005$. Washington, DC: Belden and Russonello, sponsored by Consultative Group on Biological Diversity.

Bennett, Stephen Earl. 1973. Consistency among the Public's Social Welfare Policy Attitudes in the 1960s. *American Journal of Political Science*, 17: 544–570.

Berelson, Bernard R., Paul F. Lazarsfeld, and William N. McPhee. 1954. *Voting: A Study of Opinion Formation in a Presidential Campaign*. Chicago: University of Chicago Press.

Berger, Ida E., and Ruth M. Corbin. 1992. Perceived Consumer Effectiveness and Faith in Others as Moderators of Environmentally Responsible Behaviors. *Journal of Public Policy and Marketing*, 11: 79–89.

Berinsky, Adam, and Steven Rosenstone. 1996. Evaluation of Environmental Policy Items on the 1995 NES Pilot Study. *1995 Pilot Study Reports*. Ann Arbor, MI: National Election Studies. ⟨http://www.umich.edu/~nes/resources/psreport/abs/95a.htm⟩.

Bernstein, Sharon. 1992. Planet Hollywood: The Hits and Misses of Television's Environmental Activism. *Los Angeles Times*, February 16, 4.

Berry, William D., and Mitchell S. Sanders. 2000. *Understanding Multivariate Research: A Primer for Beginning Social Scientists*. Boulder, CO: Westview Press.

Bishop, George F., Alfred F. Tuchfarber, and Robert W. Oldendick. 1986. Opinions on Fictitious Issues: The Pressure to Answer Survey Questions. *Public Opinion Quarterly*, 50: 240–250.

Bishop, Richard. 1982. Option Value: An Exposition and Extension. *Land Economics*, 58: 1–15.

Black, J. S. 1978. Attitudinal, Normative, and Economic Factors in Early Response to an Energy-Use Field Experiment. Ph.D. dissertation, University of Wisconsin, Madison.

Blalock, Hubert M., Jr. 1965. Some Implications of Random Measurement Error for Causal Inferences. *American Journal of Sociology*, 71: 37–47.

Blalock, Hubert M., Jr. 1969. Multiple Indicators and the Causal Approach to Measurement Error. *American Journal of Sociology*, 75: 264–272.

Blalock, Hubert M., Jr. 1970. A Causal Approach to Nonrandom Measurement Errors. *American Political Science Review*, 64: 1099–1111.

Blocker, T. Jean, and Douglas Lee Eckberg. 1989. Environmental Issues as Women's Issues: General Concerns and Local Hazards. *Social Science Quarterly*, 70(3): 586–593.

Blocker, T. Jean, and Douglas Lee Eckberg. 1997. Gender and Environmentalism: Results from the 1993 General Social Survey. *Social Science Quarterly*, 78: 841–858.

Bollen, Kenneth A. 1989. *Structural Equations with Latent Variables*. New York: Wiley.

Bollen, Kenneth A., and J. Scott Long. 1993. *Testing Structural Equation Models*. Newbury Park, CA: Sage.

Bone, Hugh A., and Robert C. Benedict. 1975. Perspectives on Direct Legislation: Washington State's Experience, 1914–1973. *Western Political Quarterly*, 28: 30–51.

Bord, Richard J., and Robert E. O'Connor. 1997. The Gender Gap in Environmental Attitudes: The Case of Perceived Vulnerability to Risk. *Social Science Quarterly*, 78: 830–840.

Borosage, Robert L., and Stanley B. Greenberg. 1997. Why Did Clinton Win? *The American Prospect*, no. 31: 17–21. ⟨http://epn.org/prospect/31/31-cnt.html⟩.

Bosso, Christopher J. 1994. After the Movement: Environmental Activism in the 1990s. In Norman J. Vig and Michael E. Kraft, eds., *Environmental Policy in the 1990s* (2nd ed., 31–50). Washington, DC: CQ Press.

Boudreaux, Donald J., Roger E. Meiners, and Todd J. Zywicki. 1999. Talk Is Cheap: The Existence Value Fallacy. *Environmental Law*, 29: 765.

Bowler, Shaun, and Todd Donovan. 1994. Economic Conditions and Voting on Ballot Propositions. *American Politics Quarterly*, 22(1): 27–40.

Bowler, Shaun, Todd Dovovan, and Trudi Happ. 1992. Ballot Propositions and Information Costs: Direct Democracy and the Fatigued Voter. *Western Political Quarterly*, 45: 559–568.

Bowman, Karlyn. 1998. Polluted Polling on Global Warming. *AEI on the Issues* (March). ⟨http://www.aei.org/oti/oti8852.htm⟩.

Brady, Henry E., Sidney Verba, and Kay Lehman Scholzman. 1995. Beyond SES: A Resource Model of Political Participation. *American Political Science Review*, 89(2): 271–285.

Bragdon, Peter, and Beth Donovan. 1990. Voters' Concerns Are Turning the Political Agenda Green. *Congressional Quarterly Weekly Report* (January 20): 186–187.

Braile, Robert. 2000. Book Review: The Land That Could Be. *Boston Globe*, May 14, Books-3.

Breyer, Stephen. 1993. *Breaking the Vicious Circle: Toward Effective Risk Regulation.* Cambridge, MA: Harvard University Press.

Broder, David. 1990. Beyond Folk Songs and Flowers. *Washington Post*, April 22, B7.

Brodsky, David M., and Edward Thompson III. 1993. Ethos, Public Choice, and Referendum Voting. *Social Science Quarterly*, 74(2): 286–299.

Brody, Charles J. 1984. Differences by Sex in Support for Nuclear Power. *Social Forces*, 63(1): 209–228.

Brody, Richard A., and Benjamin I. Page. 1972. Comment: The Assessment of Issue Voting. *American Political Science Review*, 66: 450–458.

Bryce, James. 1981. The Nature of Public Opinion. In Morris Janowitz and Paul M. Hirsch, eds., *Reader in Public Opinion and Mass Communication* (3rd ed., 3–9). New York: Free Press.

Buchanan, James M. 1954. Individual Choice in Voting and the Market. *Journal of Political Economy*, 62: 334–343.

Buchanan, James M., and Gordon Tullock. 1962. *The Calculus of Consent: Logical Foundations of Constitutional Democracy.* Ann Arbor: University of Michigan Press.

Bullard, Robert D. 1990. *Dumping in Dixie: Race, Class, and Environmental Quality.* Boulder, CO: Westview Press.

Burns, Nancy, Donald R. Kinder, Steven J. Rosenstone, Virginia Sapiro, and the National Election Studies. 2001. *National Election Studies, 2000: Pre- / Post-Election Study* [dataset]. Ann Arbor: University of Michigan, Center for Political Studies.

Butler, David, and Austin Ranney, eds. 1978. *Referendums: A Comparative Study of Practice and Theory.* Washington, DC: American Enterprise Institute.

Butler, David, and Austin Ranney, eds. 1994. *Referendums around the World: The Growing Use of Direct Democracy.* Washington, DC: American Enterprise Institute.

Buttel, Frederick H. 1975. Class Conflict, Environmental Conflict, and the Environmental Movement: The Social Bases of Mass Environmental Beliefs, 1968–1974. Ph.D. dissertation, University of Wisconsin, Madison.

Buttel, Frederick H. 1979. Age and Environmental Concern: A Multivariate Analysis. *Youth and Society*, 10: 237–256.

Buttel, Frederick H., and William L. Flinn. 1976. Environmental Politics: The Structuring of Partisan and Ideological Cleavages in Mass Environmental Attitudes. *Sociological Quarterly*, 17: 477–490.

Buttel, Frederick H., and William L. Flinn. 1978. The Politics of Environmental Concern: The Impacts of Party Identification and Political Ideology on Environmental Attitudes. *Environment and Behavior*, 10: 17–36.

Buttel, Frederick H., and Donald E. Johnson. 1977. Dimensions of Environmental Concern: Factor Structure, Correlates Implications for Research. *Journal of Environmental Education*, 9: 49–64.

Callahan, Deb. 2000. Environment Is "Sleeper" Issue of 2000 Campaign. *Environmental Network News* (March 9). ⟨http://www.enn.com/enn-features-archive/2000/03/03092000/callahan_10788.asp⟩.

Calvert, Jerry W. 1979. The Social and Ideological Bases of Support for Environmental Legislation: An Examination of Public Attitudes and Legislative Action. *Western Political Quarterly*, 32: 327–337.

Cambridge Reports Research International. Various years. Omnibus Survey [datafile]. September 1991; September 1992; July 15–27, 1993; September 1993; July 1994, $n = 1,250$; September 1994, $n = 1,250$. Cambridge, MA: Cambridge Reports Research International.

Campbell, Angus, Philip E. Converse, Warren E. Miller, and Donald E. Stokes. 1960. *The American Voter*. Chicago: University of Chicago Press.

Carlin, Alan, Paul F. Scodari, and Don H. Garner. 1992. Environmental Investments: The Cost of Cleaning Up. *Environment*, 34: 12–27.

Carlson, Les, Stephen J. Grove, and Norman Kangun. 1993. A Content Analysis of Environmental Advertising Claims: A Matrix Method Approach. *Journal of Advertising*, 22(3): 27–39.

Carmines, Edward G., and James A. Stimson. 1980. The Two Faces of Issue Voting. *American Political Science Review*, 74: 78–91.

Carson, Patrick, and Julia Moulden. 1991. *Green Is Gold: Business Talking to Business about the Environmental Revolution*. Toronto: HarperBusiness.

Carson, Rachel. 1962. *Silent Spring*. Boston: Houghton Mifflin.

Casuso, Jorg. 1990. Black Tuesday for Big Green Backers. *Chicago Tribune*, November 11, 6.

Catton, William R., and Riley E. Dunlap. 1978. Environmental Sociology: A New Paradigm. *American Sociologist*, 13: 41–49.

CBS News/New York Times. 1997. Survey [datafile]. November 23–24, $n = 953$. New York: CBS and New York Times.

Charlton Research Company. 1997. Research Poll [datafile]. November 17–20, $n = 800$. Washington, DC: Charlton Research.

Choudhury, Askar H., Robert Hubata, and Robert D. St. Louis. 1999. Understanding Time-Series Regression Estimators. *American Statistician*, 53: 342.

Citrin, Jack, Beth Reingold, and Donald P. Green. 1990. American Identity and the Politics of Ethnic Change. *Journal of Politics*, 52: 1124–1154.

Citrin, Jack, Beth Reingold, Evelyn Walters, and Donald P. Green. 1990. The "Official English" Movement and the Symbolic Politics of Language in the United States. *Western Political Quarterly*, 43: 535–559.

Coddington, Walter. 1993. *Environmental Marketing: Positive Strategies for Reaching the Green Consumer*. New York: McGraw-Hill.

Commoner, Barry. 1971. *The Closing Circle: Nature, Man, and Technology*. New York: Knopf.

Connelly, John, 1996. The Politics of the Environment. *Atlanta Journal and Constitution*, February 4, H1.

Constantini, Edmond, and Kenneth Hanf. 1972. Environmental Concern and Lake Tahoe: A Study of Elite Perceptions, Backgrounds, and Attitudes. *Environment and Behavior*, 4: 209–242.

Converse, Philip E. 1964. The Nature of Belief Systems in Mass Publics. In D. E. Apter, ed., *Ideology and Discontent*. New York: Free Press.

Conway, M. Margaret. 1991. *Political Participation in the United States* (2nd ed.). Washington, DC: Congressional Quarterly Press.

Cooley, Richard, and Geoffrey Wandesforde-Smith, eds. 1970. *Congress and the Environment*. Seattle: University of Washington Press.

Crosby, Lawrence A., James D. Gill, and James R. Taylor. 1981. Consumer/ Voter Behavior in the Passage of the Michigan Container Law. *Journal of Marketing*, 45 (Spring): 19–32.

Daley, Patrick, and Dan O'Neill. 1991. Sad Is Too Mild a Word: Press Coverage of the *Exxon Valdez* Oil Spill. *Journal of Communication*, 41: 53.

Darcy, R., and Ian McAllister. 1990. Ballot Position Effects. *Electoral Studies*, 9: 5–17.

Davidson, Debra J., and William R. Freudenburg. 1996. Gender and Environmental Risk Concerns: A Review and Analysis of Available Research. *Environment and Behavior*, 28: 302–339.

Dawson, Bill. 1996. Poll Shows Texans Wanting to Recycle. *Houston Chronicle*, January 3, 12.

Deacon, Robert, and Perry Shapiro. 1975. Private Preference for Collective Goods Revealed through Voting on Referenda. *American Economic Review*, 65 (December): 943–955.

deHaven-Smith, Lance. 1988. Environmental Belief Systems: Public Opinion on Land-Use Regulation in Florida. *Environment and Behavior*, 20: 176–199.

deHaven-Smith, Lance. 1989. Toward a Communicative Theory of Environmental Opinion. *Environment and Behavior*, 21: 630–635.

deHaven-Smith, Lance. 1991. *Environmental Concern in Florida and the Nation*. Gainsville: University of Florida Press.

De Young, R. 1985–1986. Encouraging Environmentally Appropriate Behavior: The Role of Intrinsic Motivation. *Journal of Environmental Systems*, 15: 281–292.

De Young, R. 1986. Some Psychological Aspects of Recycling: The Structure of Conservation Satisfactions. *Environment and Behavior*, 18: 435–449.

Diekmann, Andreas, and Axel Franzen. 1999. The Wealth of Nations and Environmental Concern. *Environment and Behavior*, 31: 540.

Dietz, Thomas, Paul C. Stern, and Gergory A. Guagnono. 1998. Social Structural and Social Psychological Bases of Environmental Concern. *Environment and Behavior*, 30: 450–481.

Dillman, Don A., and James A. Christenson. 1972. The Public Value for Pollution Control. In William R. Burch, Jr., Neil H. Cheek, Jr., and Lee Taylor, eds., *Social Behavior, Natural Resources, and the Environment* (237–256). New York: Harper and Row.

Douglas, Mary T., and Aaron Wildavsky. 1982. *Risk and Culture: An Essay on the Selection of Technical and Environmental Dangers*. Berkeley: University of California Press.

Dowie, Mark. 1991–1992. American Environmentalism: A Movement Courting Irrelevance. *World Policy Journal*, 9(1): 67–92.

Dowie, Mark. 1995. *Losing Ground: American Environmentalism at the Close of the Twentieth Century*. Cambridge, MA: MIT Press.

Downs, Anthony. 1957. *An Economic Theory of Democracy*. New York: Harper.

Downs, Anthony. 1972. Up and Down with Ecology: The Issue-Attention Cycle. *The Public Interest*, 28 (Summer): 38–50.

Drum, David. 1997. Product Placement Matures into Placement of Nonprofit Causes. *Variety*, 369 (November 17): S27–S28.

Dumanoski, Dianne. 1992a. Advertising Blitz Erodes Support for Question 3. *Boston Globe*, October 24, 17.

Dumanoski, Dianne. 1992b. Environment Not Gaining Ground during Campaign. *Boston Globe*, October 4, 1.

Dunlap, Riley E. 1975. The Impact of Political Orientation on Environmental Attitudes and Action. *Environment and Behavior*, 7: 428–454.

Dunlap, Riley E. 1987. Polls, Pollution, and Politics Revisited: Public Opinion on the Environment in the Reagan Era. *Environment*, 29 (July–August): 6–11, 32–37.

Dunlap, Riley E. 1989. Public Opinion and Environmental Policy. In James P. Lester, ed., *Environmental Politics and Policy: Theories and Evidence* (87–135). Durham, NC: Duke University Press.

Dunlap, Riley E. 1991a. Public Opinion in the 1980s: Clear Consensus, Ambiguous Commitment. *Environment*, 33(8): 10–15, 32–37.

Dunlap, Riley E. 1991b. Trends in Public Opinion Toward Environmental Issues: 1965–1990. *Society and Natural Resources*, 4: 285–312.

Dunlap, Riley E. 1998. Lay Perceptions of Global Risk: Public Views of Global Warming in Cross-National Context. *International Sociology*, 13: 473–498.

Dunlap, Riley E. 2000. Americans Have Positive Image of the Environmental Movement. *Gallup Poll Monthly*, 415 (April): 19–25.

Dunlap, Riley E., and Michael Patrick Allen. 1976. Partisan Differences on Environmental Issues: A Congressional Roll-Call Analysis. *Western Political Quarterly*, 29: 384–397.

Dunlap, Riley E., and Curtis E. Beus. 1992. Understanding Public Concerns about Pesticides: An Empirical Examination. *Journal of Consumer Affairs*, 26: 418–438.

Dunlap, Riley E., and Don A. Dillman. 1976. Decline in Public Support for Environmental Protection: Evidence from a 1970–1974 Panel Study. *Rural Sociology*, 41: 382–390.

Dunlap, Riley E., and Richard Gale. 1974. Party Membership and Environmental Politics: A Legislative Roll-Call Analysis. *Social Science Quarterly*, 55: 670–690.

Dunlap, Riley E., J. Keith Grieneeks, and Milton Rokeach. 1983. Human Values and Pro-Environmental Behavior. In W. David Conn, ed., *Energy and Material Resources: Attitudes, Values, and Public Policy* (145–168). Boulder, CO: Westview Press.

Dunlap, Riley E., and Angela G. Mertig. 1991. The Evolution of the U.S. Environmental Movement from 1970 to 1990: An Overview. *Society and Natural Resources*, 4: 209–218.

Dunlap, Riley E., and Angela G. Mertig. 1995. Global Concern for the Environment: Is Affluence a Prerequisite? *Journal of Social Issues*, 31: 121–137.

Dunlap, Riley E., and Rik Scarce. 1991. The Polls: Environmental Problems and Protection. *Public Opinion Quarterly*, 55: 651–672.

Dunlap, Riley E., and Kent D. Van Liere. 1977. Further Evidence of Declining Public Concern with Environmental Problems: A Research Note. *Western Sociological Review*, 8: 108–112.

Dunlap, Riley E., and Kent D. Van Liere. 1978. The "New Environmental Paradigm." *Journal of Environmental Education*, 4: 10–19.

Dunlap, Riley E., and Kent D. Van Liere. 1984. Commitment to the Dominant Social Paradigm and Concern for Environmental Quality. *Social Science Quarterly*, 65(4): 1013–1028.

Dunlap, Riley E., Kent D. Van Liere, and Don A. Dillman. 1979. Evidence of Decline in Public Concern with Environmental Quality: A Reply. *Rural Sociology*, 44: 204–212.

Dunlap, Riley E., Kent D. Van Liere, Angela G. Mertig, and Robert Emmet Jones. 2000. Measuring Endorsement of the New Ecological Paradigm: A Revised NEP Scale. *Journal of Social Issues*, 56: 425–442.

Durr, Robert H. 1993. What Moves Policy Sentiment? *American Political Science Review*, 87: 158–170.

Dwyer, John F. 1980. Economic Benefits of Wildlife-Related Recreation Experiences. In William W. Shaw and Ervin H. Zube, eds., *Wildlife Values* (62–69). Tucson: Center for the Assessment of Noncommodity Natural Resource Values.

Easterbrook, Gregg. 1995. *A Moment on the Earth: The Coming Age of Environmental Optimism.* New York: Viking.

Easterbrook, Gregg. 1999. Surburban Myth. *New Republic* (March 15): 18.

Eddy, Mark. 1998. Wilderness Expansion Backed; 80 percent Favor Land Protection in Colorado Poll. *Denver Post*, April 8, B6.

Eddy, Mark. 1999. Loveland Sets Bar High for Recycling: "Pay as You Throw" Plan Gaining Fame. *Denver Post*, March 7, A17.

Ellen, Pam Scholder, Joshua Lyle Wiener, and Cathy Cobb-Walgren. 1991. The Role of Perceived Consumer Effectiveness in Motivating Environmentally Conscious Behaviors. *Journal of Public Policy and Marketing*, 10(2): 102–117.

Elliott, Euel, James L. Regens, and Barry J. Seldon. 1995. Exploring Variation in Public Support for Environmental Protection. *Social Science Quarterly*, 76: 41–52.

Elliott, Euel, Barry J. Seldon, and James Regens. 1997. Political and Economic Determinants of Individuals' Support for Environmental Spending. *Journal of Environmental Management*, 51: 15–27.

Ellis, Richard J., and Fred Thompson. 1997. Culture and the Environment in the Pacific Northwest. *American Political Science Review*, 91: 885–897.

Enelow, James M., and Melvin J. Hinich. 1984. *The Spatial Theory of Voting: An Introduction.* Cambridge: Cambridge University Press.

Environmental Protection: Is the Public Willing to Pay? 1995. *EPA Journal* (Winter): 15.

Environmental Protection Agency. 1990. *Environmental Investments: The Cost of a Clean Environment.* Washington, DC: Environmental Protection Agency.

Epstein, Tom. 1986. Campaign to Defeat Proposition 65. *Los Angeles Times*, November 27, II-6.

Erikson, Robert S., Norman R. Luttbeg, and Kent L. Tedin. 1991. *American Public Opinion* (4th ed.). New York: Macmillan.

Erskine, Hazel. 1972. The Polls: Pollution and Its Costs. *Public Opinion Quarterly*, 35: 120–135.

Etzioni, Amitai. 1988. *The Moral Dimension.* New York: Free Press.

Fenly, Leigh. 1990. "New Books Tell Children about Environment and Their Future." *San Diego Union-Tribune*, April 15, Books-5.

Fenno, Richard F. 1978. *Home Style: House Members in Their Districts.* Boston: Little, Brown.

Field Institute. 1986. Field (California) Poll [datafile]. July 24–August 4, 1986 (USCA 86-04), $n = 1,028$; September 24–October 2, 1986, $n = 1,023$ (USCA

86-05); October 29–October 30, 1986, $n = 701$ (USCA 86-06). San Francisco: Field Institute, archived at University of California and at Roper Research Center.

Finger, Matthias. 1994. From Knowledge to Action? Exploring the Relationships between Environmental Experiences, Learning, and Behavior. *Journal of Social Issues*, 50: 141–160.

Firebaugh, Glenn, and Kenneth E. Davis. 1988. Trends in Anti-Black Prejudice, 1972–1984: Region and Cohort Effects. *American Journal of Sociology*, 94: 251–272.

Fischer, David W. 1975. Willingness to Pay as a Behavioral Criterion for Environmental Decision-making. *Journal of Environmental Management*, 3: 29–41.

Fishbein, Martin, and Icek Ajzen. 1975. *Belief, Attitude, Intention, and Behavior: An Introduction to Theory and Research*. Reading, MA: Addison-Wesley.

Fried, Amy. 1998. U.S. Environmental Interest Groups and the Promotion of Environmental Values: The Resounding Success and Failure of Earth Day. *Environmental Politics*, 7: 1–2.

Friedman, Monroe. 1995. On Promoting a Sustainable Future through Consumer Activism. *Journal of Social Issues*, 51: 197–215.

Funkhouser, G. R. 1973. The Issues of the Sixties: An Exploratory Study in the Dynamics of Public Opinion. *Public Opinion Quarterly*, 33: 62–75.

Furman, Andrzej, and Oguz Erdur. 1999. Are Good Citizens Environmentalists? *Human Ecology: An Interdisciplinary Journal*, 27: 181–188.

Gallup Organization. Various years. The Gallup Poll [datafile]. April 5–8, 1990, $n = 1,223$; April 3–9, 2000, $n = 1,004$; May 23–24, 2000, $n = 1,032$; March 5–7, 2001, $n = 1,060$. Princeton, NJ: Gallup.

Garland, Susan B. 1996. Come Winter, a Greener Hill? *Business Week* (April 15): 66.

Gelernter, David. 1994. The Immorality of Environmentalism. *City Journal* (Autumn): 14.

Geller, E. Scott, Richard A. Winett, and Peter B. Everett. 1982. *Preserving the Environment: Strategies for Behavioral Change*. New York: Pergamon.

Geller, Jack M., and Paul Lasley. 1985. The New Environmental Paradigm Scale: A Reexamination. *Journal of Environmental Education*, 10: 9–12.

Gerlak, Andrea K., and Susan M. Natali. 1993. *Taking the Initiative, II*. Washington, DC: Americans for the Environment.

Gerstenzang, James. 1997. Survey Bolsters Global Warming Fight. *Los Angeles Times*, November 21, A4.

Gillroy, John M., and Robert Y. Shapiro. 1986. The Polls: Environmental Protection. *Public Opinion Quarterly*, 50: 270–279.

Gilmore, Susan. 1990. Nation's Voters Decide to Keep Their Wallets Green: Environmental Defeats Blamed on Economy. *Seattle Times*, November 8, A5.

Glenn, Norval B. 1975. Trend Studies with Available Survey Data: Opportunities and Pitfalls. In *Survey Data for Trend Analysis: An Index to Repeated Questions on U.S. National Surveys Held by the Roper Public Opinion Research Center* (6–48). Williamstown, MA: Roper Center.

Goodrich, Chris. 1995. Just a Few Words of Optimism from Green Enemy No. 1. *Los Angeles Times*, June 16, E6.

GOP Hears Nature's Call: The Party Scrambles on Environmental Policy. 1996. *Time*, 146 (March 4), 57.

Gore, Al. 1992. *Earth in the Balance: Ecology and the Human Spirit*. Boston: Houghton Mifflin.

Gottlieb, Robert. 1993. *Forcing the Spring: The Transformation of the American Environmental Movement*. Washington, DC: Island Press.

Gottlieb, Robert, and Helen Ingram. 1988. The New Environmentalists. *The Progressive*, 52: 14–15.

Graham, Virginia. 1978. *A Compilation of Statewide Initiative Proposals Appearing on Ballots through 1976*. Washington, DC: Congressional Research Service, Library of Congress.

Gray-Lee, Jason W., Debra L. Scammon, and Robert N. Mayer. 1994. Review of Legal Standards for Environmental Marketing Claims. *Journal of Public Policy and Marketing*, 13(1): 155–159.

Green Choices, Hard Choices: The Environmental Risks That Frighten People Most Rarely Matter Most. 1991. *The Economist*, 318 (March 16): 11.

Green, Donald Philip. 1988. On the Dimensionality of Public Sentiment toward Partisan and Ideological Groups. *American Journal of Political Science*, 32: 758–780.

Green, Donald Philip. 1992. The Price Elasticity of Mass Preferences. *American Political Science Review*, 86 (March): 128–148.

Green, Donald Philip, and Irene V. Blair. 1995. Framing and the Price Elasticity of Private and Public Goods. *Journal of Consumer Psychology*, 4(1): 1–32.

Green, Donald Philip, and Jack Citrin. 1994. Measurement Error and the Structure of Attitudes: Are Positive and Negative Judgments Opposites? *American Journal of Political Science*, 38: 256–281.

Green, Donald Philip, Daniel Kahneman, and Howard Kunreuther. 1994. How the Scope and Method of Public Funding Affect Willingness to Pay for Public Goods. *Public Opinion Quarterly*, 58: 49–67.

Guber, Deborah Lynn. 1996. Environmental Concern and the Dimensionality Problem: A New Approach to an Old Predicament. *Social Science Quarterly*, 77(3): 644–662.

Gup, Ted. 1990. Owl vs. Man. *Time* (June 25): 56–62.

Halverson, Richard. 1991. Big Three Take High Road on Environmental Front. *Discount Store News*, 30 (March 18): 72.

Hardin, Garrett. 1968. The Tragedy of the Commons. *Science* (December 13): 1243–1248.

Harris, Jack. 1993. Experts in Everything and Nothing. *New Scientist*, 138: 45.

Harris, John F., and Ellen Nakashima. 2000. Gore's Greenness Fades: Political Caution Has Tempered His Environmentalism. *Washington Post*, February 28, A1.

Harris, Louis, and Associates, Inc. 1993. The Harris Poll [datafile]. April 28– May 4, 1993, $n = 1,252$.

Harris, Louis, and Associates, Inc. 1989. *Public and Leadership Attitudes to the Environment in Four Continents: A Report of a Survey in Sixteen Countries.* New York: Louis Harris and Associates, for the United Nations Environmental Programme.

Hayduk, Leslie A. 1987. *Structural Equation Modeling with LISREL: Essentials and Advances.* Baltimore, MD: Johns Hopkins University Press.

Hays, Samuel P. 1987. *Beauty, Health, and Permanence: Environmental Politics in the United States, 1955–1985.* Cambridge: Cambridge University Press.

Helvarg, David. 1994. *The War against the Greens: The "Wise-Use" Movement, the New Right, and Anti-Environmental Violence.* San Francisco: Sierra Club Books.

Herndl, Carl G., and Stuart C. Brown. 1996. *Green Culture: Environmental Rhetoric in America.* Madison: University of Wisconsin Press.

Hershkowitz, Allen. 1991. Green vs. Greenbacks. *Advertising Age* (January 29): GR10–GR11.

Hetherington, Mark. 1996. The Media's Role in Forming Voters' National Economic Evaluations in 1992. *American Journal of Political Science*, 40: 372–395.

Hibbing, John R., and Elizabeth Theiss-Morse. 1995. *Congress as Public Enemy: Public Attitudes toward American Political Institutions.* Cambridge: Cambridge University Press.

Holmes, Thomas. P. 1990. Self-Interest, Altruism, and Health-Risk Reduction: An Economic Analysis of Voting Behavior. *Land Economics*, 66(2): 140–149.

Holusha, John. 1990. Talking Deals: Unusual Alliance for McDonald's. *New York Times*, August 9, D2.

Honnold, Julie A. 1984. Age and Environmental Concern: Some Specification of Effects. *Journal of Environmental Education*, 16: 4–9.

Hopper, Joseph R., and Joyce McCarl Nielsen. 1991. Recycling as Altruistic Behavior: Normative and Behavioral Strategies to Expand Participation in a Community Recycling Program. *Environment and Behavior*, 23: 195–220.

Hornback, Kenneth E. 1974. Orbits of Opinion: The Role of Age in the Environmental Movement's Attentive Public, 1968–1972. Ph.D. dissertation, Michigan State University, East Lansing.

Hotelling, H. 1947. Letter to Director of National Park Service. In *An Economic Study of the Monetary Evaluation of Recreation in the National Parks*. U.S. Department of the Interior, National Park Service and Recreational Planning Division.

How Green Is Al Gore? 2000. *The Economist* (April 22): 30.

Huber, Peter. 1996. How to Love Butterflies and Be a Conservative. *Forbes*, 157 (April 22): 216–217.

Hume, Scott. 1991. Consumer Doubletalk Makes Companies Wary: "Preferring" to Buy Green Isn't Proven by Sales. *Advertising Age* (October 28): GR4.

Hungerford, Don, and Barbara Lewis. 1997. Is Environmental Education Scaring Our Children to Death? *NEA Today*, 16: 43.

Immerwahr, John. 1999. *Waiting for a Signal: Public Attitudes toward Global Warming, the Environment and Geophysical Research*. American Geophysical Union, Electronic Publication. ⟨http://www.agu.org/sci_soc/attitude_study.html⟩.

Inglehart, Ronald. 1981. Post-Materialism in an Environment of Insecurity. *American Political Science Review*, 75: 880–899.

Jacobs, Paul. 1986. Prop. 65: Toxics Calamity or Legal Catalyst? *Los Angeles Times*, October 13, 1.

Janowitz, Morris, and Paul M. Hirsch, eds. 1981. *Reader in Public Opinion and Mass Communication* (3rd ed., 3–9). New York: Free Press.

Jensen, Elizabeth. 2000. DiCaprio and "Planet Earth" Rank Fourth in Time Slot. *Los Angeles Times*, April 24, F2.

Johnson, John Mark. 1990. Citizens Initiate Ballot Measures. *Environment*, 32(7): 4–5, 43–45.

Jones, Robert Emmet, and Lewis F. Carter. 1994. Concern for the Environment among Black Americans: An Assessment of Common Assumptions. *Social Science Quarterly*, 75(3): 560–579.

Jones, Robert Emmet, and Riley E. Dunlap. 1992. The Social Bases of Environmental Concern: Have They Changed over Time? *Rural Sociology*, 57: 28–47.

Jöreskog, Karl G., and Dag Sörbom. 1989. *LISREL 7: A Guide to the Program and Applications*. Chicago: SPSS.

Kachigan, Sam Kash. 1991. *Multivariate Statistical Analysis: A Conceptual Introduction* (2nd ed.). New York: Radius Press.

Kahn, M. E., and J. G. Matsusaka. 1997. Demand for Environmental Goods: Evidence from Voting Patterns on California Initiatives. *Journal of Law and Economics*, 40: 137–173.

Kamieniecki, Sheldon. 1995. Political Parties and Environmental Policy. In James P. Lester, ed., *Environmental Politics and Policy: Theories and Evidence* (2nd ed., 146–167). Durham, NC: Duke University Press.

Kanagy, Conrad L., Craig R. Humphrey, and Glenn Firebaugh. 1994. Surging Environmentalism: Changing Public Opinion or Changing Publics? *Social Science Quarterly*, 75: 804–819.

Karlberg, Michael. 1997. News and Conflict: How Adversarial News Frames Limit Public Understanding of Environmental Issues. *Alternatives Journal*, 23: 22–28.

Kasper, Raphael G. 1980. Perceptions of Risk and Their Effects on Decision Making. In Richard C. Schwing and Walter A. Albers Jr., eds., *Societal Risk Assessment: How Safe Is Safe Enough?* (71–80). New York: Plenum Press.

Kates, Robert W. 2000. Has the Environment Improved? *Environment*, 42: FC.

Katosh, John P., and Michael W. Traugott. 1981. The Consequences of Validated and Self-reported Voting Measures. *Public Opinion Quarterly*, 45: 519–535.

Keeter, Scott. 1984. Problematical Pollution Polls: Validity in the Measurement of Public Opinion on Environmental Issues. *Political Methodology*, 10: 267–291.

Kelman, Steven. 1981. *What Price Incentives? Economists and the Environment*. Boston: Auburn House.

Kempton, Willett, James S. Boster, and Jennifer A. Hartley. 1995. *Environmental Values in American Culture*. Cambridge, MA: MIT Press.

Kennedy, Peter. 1993. *A Guide to Econometrics* (3rd ed.). Cambridge, MA: MIT Press.

Kerlinger, Fred N. 1979. *Behavioral Research: A Conceptual Approach*. Dallas: Holt, Rinehart, and Winston.

Key, V. O., Jr. 1961. *Public Opinion and American Democracy*. New York: Knopf.

Kinder, Donald R., and D. Roderick Kiewiet. 1981. Sociotropic Politics: The American Case. *British Journal of Political Science*, 11: 129–161.

Kinder, Donald R., and Lynn M. Sanders. 1990. Mimicking Political Debate with Survey Questions: The Case of White Opinion on Affirmative Action for Blacks. *Social Cognition*, 8: 73–103.

King, Gary. 1989. *Unifying Political Methodology: The Likelihood Theory of Statistic Inference*. Cambridge: Cambridge University Press:

Kingdon, John W. 1984. *Agendas, Alternatives, and Public Policies*. Boston: Little, Brown.

Kinnear, Thomas C., James R. Taylor, and Sadrudin A. Ahmed. 1974. Ecologically Concerned Consumers: Who Are They? *Journal of Marketing*, 38: 20–24.

Klein, Gil. 1995. Environment Challenges Public Mood. *Tampa Tribune*, July 22, 4.

Klineberg, Stephen L., Matthew McKeever, and Bert Rothenbach. 1998. Demographic Predictors of Environmental Concern: It Does Make a Difference How It's Measured. *Social Science Quarterly*, 79: 734–753.

Kluegel, James R., and Eliot R. Smith. 1982. Whites' Beliefs about Blacks' Opportunity. *American Sociological Review*, 47: 518–532.

Knickerbocker, Brad. 1995a. Americans Go "Lite Green." *Christian Science Monitor*, April 18, 1.

Knickerbocker, Brad. 1995b. Is It Wrong to Be Optimistic about the Environment? *Christian Science Monitor*, June 13, 14.

Knopman, Debra S., Megan M. Susman, and March K. Landy. 1999. Civic Environmentalism: Tackling Tough Land-Use Problems with Innovative Governance. *Environment*, 41: 24.

Kriz, Margaret E. 1995. The Green Card. *National Journal*, 27(37): 2262.

Kriz, Margaret. 1996. Slinging Earth. *National Journal*, 28(17): 958.

Kuhn, Richard G., and Edgar L. Jackson. 1989. Stability of Factor Structures in the Measurement of Public Environmental Attitudes. *Journal of Environmental Education*, 20: 27–32.

Lacayo, Richard. 1990. Green Ballots vs. Greenbacks. *Time* (November 19): 44.

Ladd, Everett Carll. 1979. The New Lines Are Drawn: Class and Ideology in America. *Public Opinion* (July–August): 52–53.

Ladd, Everett Carll. 1982. Clearing the Air: Public Opinion and Public Policy on the Environment. *Public Opinion* (August–September): 16–20.

Ladd, Everett Carll. 1990. What Do Americans Really Think about the Environment? *Public Perspective*, 1(4): 11.

Ladd, Everett Carll, and Karlyn H. Bowman. 1995. *Attitudes toward the Environment: Twenty-Five Years after Earth Day*. Washington, DC: American Enterprise Institute Press.

Lake, Laura M. 1983. The Environmental Mandate: Activists and the Electorate. *Political Science Quarterly*, 98(2): 215–233.

La Trobe, Helen L., and Tim G. Acott. 2000. A Modified NEP/DSP Environmental Attitudes Scale. *Journal of Environmental Education*, 32(1): 12.

Lau, Richard R. 1985. Two Explanations for Negativity Effects in Political Behavior. *American Journal of Political Science*, 29: 119–138.

Lauter, David, and Douglas Jehl. 1992. Bush, Clinton Clash over Jobs, Environment. *Los Angeles Times*, September 15, A1.

LaVally, Rebecca. 1987. Proposition 65 Opponents Outspent Successful Foes. United Press International, February 5.

Leaversuch, Robert D., ed. 1992. Massachusetts Votes Down Drastic Measure to Reduce Package Waste. *Modern Plastics*, 69(13): 37.

Lee, Eugene. 1978. California. In David Butler and Austin Ranney, eds., *Referendums: A Comparative Study of Practice and Theory*. Washington, DC: American Enterprise Institute.

Leroux, Kivi. 1999. Subliminal Messages. *E*, 10 (July): 14.

Lester, James P., ed. 1995. *Environmental Politics and Policy: Theories and Evidence*. Durham, NC: Duke University Press.

Levitt, Norman. 1999. *Prometheus Bedevilled: Science and the Contradictions of Contemporary Culture*. Piscataway, NJ: Rutgers University Press.

Locke, Robert. 1986a. California Oil Industry Joins Others in Wariness over Proposition 65. *Oil Daily*, September 18, 8.

Locke, Robert. 1986b. Success of Proposition 65, Approval of Restriction Cloud Election Picture. *Oil Daily*, November 20, 8.

Lomborg, Bjorn. 2001. *The Skeptical Environmentalist: Measuring the Real State of the World*. Cambridge: Cambridge University Press.

Loth, Renee. 1991. Bringing Earth Day Back down to Earth: Grass-Roots Activists Tweak "Elitist" Brethren. *Boston Globe*, April 21, A33.

Lounsbury, J. W., and L. G. Tornatzky. 1977. A Scale for Assessing Attitudes toward Environmental Quality. *Journal of Social Psychology*, 101: 299–305.

Lovett, Richard A. 1994. Proposition 65 Comes of Age. *California Journal* (November 1): 25–28.

Lowe, George D., Thomas K. Pinhey, and Michael D. Grimes. 1980. Public Support for Environmental Protection: New Evidence from National Surveys. *Pacific Sociological Review*, 23(4): 423–445.

Lowell, A. Lawrence. 1981. Public Opinion. In Morris Janowitz and Paul M. Hirsch, eds., *Reader in Public Opinion and Mass Communication* (3rd ed., 10–16). New York: Free Press.

Lowenstein, D. H. 1982. Campaign Spending and Ballot Propositions: Recent Experience, Public Choice Theory and the First Amendment. *U.C.L.A. Law Review*, 29: 505–641.

MacDonald, Stuart Elaine, George Rabinowitz, and Ola Listhaug. 1995. Political Sophistication and Models of Issues Voting. *British Journal of Political Science*, 25(4): 453–483.

MacDonald, Stuart Elaine, George Rabinowitz, and Ola Listhaug. 1998. On Attempting to Rehabilitate the Proximity Model: Sometimes the Patient Just Can't Be Helped. *Journal of Politics*, 60(3): 653–690.

MacKuen, Michael B. 1984. Reality, the Press, and Citizens' Political Agendas. In Charles F. Turner and Elizabeth Martin, eds., *Surveying Subjective Phenomena*. New York: Russell Sage Foundation.

Magleby, David B. 1984. *Direct Legislation: Voting on Ballot Propositions in the United States*. Baltimore: Johns Hopkins University Press.

Magleby, David B. 1989. Opinion Formation and Opinion Change in Ballot Proposition Campaigns. In M. Margolis and G. Mauser, eds., *Manipulating Public Opinion* (95–115). Belmont, CA: Brooks Cole.

Magleby, David B. 1994. Direct Legislation in the American States. In David Butler and Austin Ranney, eds., *Referendums around the World: The Growing Use of Direct Democracy* (218–257). Washington, DC: American Enterprise Institute.

Mainieri, Elaine, G. Barnett, Trisha R. Valdero, John B. Unipan, and Stuart Oskamp. 1997. Green Buying: The Influence of Environmental Concern on Consumer Behavior. *Journal of Social Psychology*, 137: 189–204.

Malik, Kenan. 1999. Why Science Needs Protection. *The Independent* (July 11): 13.

Malkis, A., and H. G. Grasmick. 1977. Support for the Ideology of the Environmental Movement: Tests of Alternative Hypotheses. *Western Sociological Review*, 8: 25–47.

Maloney, Michael P., Michael P. Ward, and G. Nicholas Braucht. 1975. A Revised Scale for the Measurement of Ecological Attitudes and Knowledge. *American Psychologist*, 30: 787–790.

Mantese, Joe. 1991. New Study Finds Green Confusion. *Advertising Age* (October 21): 1.

Margolis, Michael. 1977. From Confusion to Confusion: Issues and the American Voter (1956–1972). *American Political Science Review*, 71: 31–43.

Marsh, C. Paul, and James A. Christenson. 1977. Support for Economic Growth and Environmental Protection, 1973–1975. *Rural Sociology*, 42: 101–107.

Martin, Glen. 2000a. Earth Day Report Card: We Still Care, Sort of. *San Francisco Chronicle*, April 22, A1.

Martin, Glen. 2000b. Environment Is a Big Concern for Californians, Poll Shows. *San Francisco Chronicle*, June 21, A3.

Martinez-Vazquez, Jorge. 1981. Selfishness versus Public "Regardingness" in Voting Behavior. *Journal of Public Economics*, 15: 349–361.

Marttila & Kiley, Inc. 1992. Survey (MK 92110) [datafile]. February 11–13, $n = 402$. Boston: Marttila & Kiley, for Massachusetts Public Interest Research Group (Mass PIRG).

Maslow, Abraham H. 1970. *Motivation and Personality* (2nd ed.). New York: Harper and Row.

Mastio, David. 2000. The GOP's Enviro-Rut. *Policy Review* (June): 19.

Mathews, Jessica. 1990. The Big, the Green, the Political. *Washington Post*, November 16, A19.

Mathews, Jessica. 1995. Outlook Not So Gloomy. *Plain Dealer*, April 24, B9.

McCloskey, Herbert. 1964. Consensus and Ideology in American Politics. *American Political Science Review*, 48: 361–382.

McCloskey, Michael. 1987. A Second-Order Issue. *Environment*, 29: 2.

Meadows, Donella H. 2000. Looking Back at Thirty Earth Days. *St. Louis Post-Dispatch*, April 25, B7.

Meadows, Donella H., Dennis L. Meadows, Jorgens Randers, and William W. Behrens III. 1972. *The Limits to Growth*. New York: Universe Books.

Mellman Group. 1997. State of the Climate [datafile]. August 11–14, $n = 800$. Washington, DC: Mellman, for World Wildlife Fund.

Mertig, Angela G., and Riley E. Dunlap. 1995. Public Approval of Environmental Protection and Other New Social Movement Goals in Western Europe and the United States. *International Journal of Public Opinion Research*, 7: 145–156.

Michaels, Patrick. 1992. Reporters Cry Wolf about Environment. *Insight on the News*, 8 (November 23): 20–21.

Milbrath, Lester W. 1965. *Political Participation*. Skokie, IL: Rand McNally.

Milbrath, Lester W. 1984. *Environmentalists: Vanguard for a New Society*. Albany: State University of New York Press.

Milbrath, Lester W., and M. L. Goel. 1977. *Political Participation* (2nd ed.). Chicago: Rand McNally.

Miller, Warren E., Donald R. Kinder, Steven J. Rosenstone, and the National Election Studies. 1999. *National Election Studies, 1991 Pilot Election Study* [dataset]. Ann Arbor: University of Michigan, Center for Political Studies.

Mr. Nader's Electoral Mischief. 2000. *New York Times*, October 26, A34.

Mr. Nader's Misguided Crusade. 2000. *New York Times*, June 30, A24.

Mitchell, Robert C. 1979. Silent Spring/Solid Majorities. *Public Opinion*, 2 (August–September): 16–20, 55.

Mitchell, Robert C. 1984. Public Opinion and Environmental Politics in the 1970s and 1980s. In Norman J. Vig and Michael E. Kraft, eds., *Environmental Policy in the 1980s* (57–74). Washington, DC: Congressional Quarterly Press.

Mitchell, Robert C. 1990. Public Opinion and the Green Lobby: Poised for the 1990s? In Norman J. Vig and Michael E. Kraft, eds., *Environmental Policy in the 1990s* (81–88). Washington: CQ Press.

Mitchell, Robert C., and Richard T. Carson. 1989. *Using Surveys to Value Public Goods: The Contingent Valuation Method*. Baltimore: Johns Hopkins University Press, for Resources for the Future.

Mitchell, Robert C., Angela A. Mertig, and Riley E. Dunlap. 1992. Twenty Years of Environmental Mobilization: Trends among National Environmental Organizations. In Riley E. Dunlap and Angela G. Mertig, eds., *American Environmentalism: The U.S. Environmental Movement, 1970–1990* (11–26). Washington, DC: Taylor & Francis.

Mohai, Paul. 1985. Public Concern and Elite Involvement in Environmental-Conservation Issues. *Social Science Quarterly*, 66: 820–838.

Mohai, Paul. 1990. Black Environmentalism. *Social Science Quarterly*, 71(4): 744–765.

Mohai, Paul. 1992. Men, Women, and the Environment: An Examination of the Gender Gap in Environmental Concern and Activism. *Society and Natural Resources*, 5: 1–19.

Mohai, Paul, and Bunyan Bryant. 1998. Is There a "Race" Effect on Concern for Environmental Quality? *Public Opinion Quarterly*, 62: 475–477.

Mohai, Paul, and Ben W. Twight. 1987. Age and Environmentalism: An Elaboration of the Buttel Model Using National Survey Evidence. *Social Science Quarterly*, 68: 798–815.

Monroe, Kristen Renwick. 1991. John Donne's People: Explaining Differences between Rational Actors and Altruists through Cognitive Frameworks. *Journal of Politics*, 53: 394–433.

Morris, Louis A., Manoj Hastak, and Michael B. Mazis. 1995. Consumer Comprehension of Environmental Advertising and Labeling Claims. *Journal of Consumer Affairs*, 29: 328–350.

Morrison, Denton E. 1986. How and Why Environmental Consciousness Has Trickled Down. In Allan Schnaiberg, Nicholas Watts, and Klaus Zimmermann, eds., *Distributional Conflict in Environmental-Resource Policy* (187–220). New York: St. Martin's Press.

Morrison, Denton E., and Riley E. Dunlap. 1986. Environmentalism and Elitism: A Conceptual and Empirical Analysis. *Environmental Management*, 10: 581–589.

Mueller, John E. 1969. Voting on the Propositions: Ballot Patterns and Historical Trends in California. *American Political Science Review*, 63: 1197–1213.

Munton, Donald, and Linda Brady. 1970. American Public Opinion and Environmental Pollution. *The Behavioral Science Laboratory Research Report*. Columbus: Ohio State University.

National Environmental Education and Training Foundation (NEETF). Various years. *National Report Card on Environmental Attitudes, Knowledge, and Behavior*. Conducated by Roper Starch Worldwide. Washington, DC: NEETF. ⟨http://www.neetf.org⟩.

National Opinion Research Center. 1994. General Social Survey [datafile]. $n = 1,334$. Chicago: NORC, University of Chicago.

National Science Foundation (NSF). 1997. Ohio State University Survey. September 1–October 5, $n = 688$. Sponsored by Resources for the Future, Washington, DC. Arlington, VA: NSF.

NBC News/Wall Street Journal. 1997. Survey [datafile]. October 28–28, $n = 1,214$. New York.

Nelkins, Dorothy. 1981. Nuclear Power as a Feminist Issue. *Environment*, 23: 14–39.

Nicholl, Jack. 1989. Son of David and Goliath: Employing Strategy, Little Guys with Initiative Can Beat Big Bucks. *Campaigns and Elections* (March–April): 16–17.

Nie, Norman, Sidney Verba, and John R. Petrocik. 1976. *The Changing American Voter*. Cambridge, MA: Harvard University Press.

Nieves, Evelyn. 2001. Conversation: Ralph Nader. A Party Crasher's Lone Regret: That He Didn't Get More Votes. *New York Times*, February 18, D7.

Noe, Francis P., and Rob Snow. 1990. The New Environmental Paradigm and Further Scale Analysis. *Journal of Environmental Education*, 21: 20–26.

Ogden, D. M., Jr. 1971. The Future of the Environmental Struggle. In Roy L. Meek and John A. Straayer, eds., *The Politics of Neglect: The Environmental Crisis* (243–250). Boston: Houghton Mifflin.

Olson, Mancur, Jr. 1965. *The Logic of Collective Action: Public Goods and the Theory of Groups*. Cambridge, MA: Harvard University Press.

Oskamp, Stuart, Maura J. Harrington, Todd C. Edwards, Deborah L. Sherwood, Shawm M. Okuda, and Deborah C. Swanson. 1991. Factors Influencing Household Recycling Behavior. *Environment and Behavior*, 23 (July): 494–519.

Ostrom, Charles W., Jr. 1978. *Time Series Analysis: Regression Techniques*. Beverly Hills: Sage.

Ottman, Jacquelyn. 1992. Environmentalism Will Be the Trend of the '90s. *Marketing News* (December 7): 13.

Ourlian, Robert. 2000. Anti-Green Measures Are Riders in the Storm. *National Journal*, 32 (January 22): 243.

Paehlke, Robert C. 1989. *Environmentalism and the Future of Progressive Politics*. New Haven: Yale University Press.

Page, Benjamin I., and Robert Y. Shapiro. 1982. Changes in Americans' Policy Preferences, 1935–1979. *Public Opinion Quarterly*, 46: 24–42.

Page, Benjamin I., and Robert Y. Shapiro. 1983. Effects of Public Opinion on Policy. *American Political Science Review*, 77: 175–190.

Page, Benjamin I., Robert Y. Shapiro, and Glenn R. Dempsey. 1987. What Moves Public Opinion? *American Political Science Review*, 81: 23–44.

Paige, Sean. 1998. The "Greening" of Government. *Insight on the News* (December 14): 16.

Parlour, J. W., and S. Schatzow. 1978. The Mass Media and Public Concern for Environmental Problems in Canada, 1960–1972. *International Journal of Environmental Studies*, 13: 9–17.

Peffley, Mark, Stanley Feldman, and Lee Sigelman. 1987. Economic Conditions and Party Competence: Processes of Belief Revision. *Journal of Politics*, 49: 100–121.

Pew Research Center for the People and the Press. 1997. *November 1997 News Interest Index* [datafile]. November 12–16, $n = 1,200$. Washington, DC: Pew Research Center.

Pierce, John C., and Nicholas P. Lovrich Jr. 1980. Belief Systems Concerning the Environment: The General Public, Attentive Publics, and State Legislators. *Political Behavior*, 2: 259–286.

Pierce, John C., Nicholas P. Lovrich, and T. Tsurutani. 1987. Culture, Politics and Mass Publics: Traditional and Modern Supporters of the New Environmental Paradigm in Japan and the United States. *Journal of Politics*, 49: 54–79.

Pirages, Dennis C., and Paul R. Ehrlich. 1974. *Ark II: Social Response to Environmental Imperatives*. San Francisco: Freeman.

Pomper, Gerald M. 1972. From Confusion to Clarity: Issues and American Voters, 1956–1968. *American Political Science Review*, 66: 415–428.

Pomper, Gerald M. 1975. *Voters' Choice: Varieties of American Electoral Behavior*. New York: Dodd, Mead.

Presser, Stanley. 1990. Can Changes in Context Reduce Vote Overreporting in Surveys? *Public Opinion Quarterly*, 54: 587.

Princeton Survey Research Associates (PSRA). Various years. Survey [datafile]. November 13–14, 1997, $n = 752$; April 13–14, 2000, $n = 752$ (conducted for *Newsweek*). Princeton, NJ: PSRA.

Purdy, Jedediah. 2000a. Planet Bush, Planet Gore. *The American Prospect*, 11 (October 20): 34.

Purdy, Jedediah. 2000b. Shades of Green. *The American Prospect*, 11 (January 3): 6.

Rabinowitz, George, and Stuart Elaine Macdonald. 1989. A Directional Theory of Issue Voting. *American Political Science Review*, 83: 93–121.

Rabinowitz, George, James W. Prothro, and William Jacoby. 1982. Salience as a Factor in the Impact of Issues on Candidate Elections. *Journal of Politics*, 44: 41–63.

Ranney, Austin. 1978. The United States of America. In David Butler and Austin Ranney, eds., *Referendums: A Comparative Study of Practice and Theory* (71–72). Washington, DC: American Enterprise Institute.

Rauber, Paul. 1991. Losing the Initiative? *Sierra*, 76: 20–24.

Rauch, Jonathan. 2000. There's Smog in the Air, But It Isn't All Pollution. *Washington Post*, April 30, B01.

Ridenour, David A. 1996. The Mouse That Squeaked. *National Policy Analysis* (November). ⟨http://www.nationacenter.inter.net/npa154.htm⟩.

Ridgeway, James. 1998. It Isn't Easy Voting Green. *Audubon*, 100(5): 144–145.

Ridley, Matt, and Bobbi S. Low. 1993. Can Selfishness Save the Environment? *Atlantic Monthly*, 272 (September): 76–86.

Roberts, James A., and Donald R. Bacon. 1997. Exploring the Subtle Relationships between Environmental Concern and Ecologically Conscious Consumer Behavior. *Journal of Business Research*, 40(1): 79–89.

Rohrschneider, Robert. 1988. Citizen's Attitudes Toward Environmental Issues: Selfish or Selfless? *Comparative Political Studies*, 21: 347–367.

Roper Center for Public Opinion Research. Various years. Survey [datafile]. Storrs, CT: Roper Center, University of Connecticut.

Rosenbaum, Walter A. 1977. *The Politics of Environmental Concern* (2nd ed.). New York: Praeger.

Rosenbaum, Walter A. 1991. *Environmental Politics and Policy* (2nd ed.). Washington, DC: Congressional Quarterly Press.

Rosenstone, Steven J., Donald R. Kinder, Warren E. Miller, and the National Election Studies. 1998. *National Election Studies, 1996: Pre- and Post-Election Study* [dataset], 3rd release. Ann Arbor: University of Michigan, Center for Political Studies.

Rosenstone, Steven J., Warren E. Miller, Donald R. Kinder, and the National Election Studies. 1999. *National Election Studies, 1995 Pilot Election Study* [datafile]. Ann Arbor: University of Michigan, Center for Political Studies.

Rothman, Hal K. 1998. *The Greening of a Nation? Environmentalism in the United States since 1945*. Orlando, FL: Harcourt Brace College.

Rourke, John T., Richard P. Hiskes, and Cyrus Ernesto Zirakzadeh. 1992. *Direct Democracy and International Politics: Deciding International Issues through Referendums*. Boulder: Rienner.

Ruben, Barbara. 1994. Back Talk. *Environmental Action*, 25: 11–16.

Rubin, Charles T. 1994. *The Green Crusade: Rethinking the Roots of Environmentalism*. Lanham, MD: Rowman & Littlefield.

Russell, Cristine. 1989. Warnings, Warnings Everywhere. *Washington Post*, May 23, Z12, Z14.

Russell, Sabin. 1990. Corporations Going for the Green. *San Francisco Chronicle*, April 9, D1.

Ryder, Norman B. 1965. The Cohort as a Concept in the Study of Social Change. *American Sociological Review*, 30: 843–861.

Saad Lydia, and Riley E. Dunlap. 2000. Americans Are Environmentally Friendly, but Issue Not Seen as Urgent Problem. *Gallup Poll Monthly*, 415 (April): 12–18.

Saffire, David. 1995. Has the Time Come for Deposit Legislation? *Beverage Industry*, 86: 25.

Sagoff, Mark. 1992. The Great Environmental Awakening. *The American Prospect* (Spring): 39–47.

Sale, Kirkpatrick, 1993, *The Green Revolution: The American Environmental Movement, 1962–1992*. New York: Hill and Wang.

Samdahl, Diane M., and Robert Robertson. 1989. Social Determinants of Environmental Concern: Specification and Test of the Model. *Environment and Behavior*, 21 (January): 57–81.

Sanera, Michael, and Jane S. Shaw. 1999. *Facts, Not Fear: Teaching Children about the Environment*. Washington, DC: Regnery.

Scammon, Debra L., and Robert N. Mayer. 1995. Agency Review of Environmental Marketing Claims: Case-by-Case Decomposition of the Issues. *Journal of Advertising*, 24(2): 33–43.

Schechter, M., M. Kim, and L. Golan. 1989. Valuing a Public Good: Direct and Indirect Valuation Approaches to the Measurement of the Benefits from Pollution Abatement. In H. Folmer, and E. ver Ierland, eds., *Valuation Methods and Policy Making in Environmental Economics*. Amsterdam: Elsevier.

Schneider, William. 1990. Everybody's An Environmentalist Now. *National Journal* (April 28): 1062.

Schroeder, Christopher H. 2000. Clear Consensus, Ambiguous Commitment. *Michigan Law Review*, 98: 1876–1915.

Schrum, L. J., Tina M. Lowrey, and John A. McCarty. 1994. Recycling as a Marketing Problem: A Framework for Strategy Development. *Psychology and Marketing*, 11: 393–416.

Schuhwerk, Melody E., and Roxanne Lefkoff-Hagius. 1995. Green or Non-Green? Does Type of Appeal Matter When Advertising a Green Product? *Journal of Advertising*, 24(2): 45–54.

Schultze, Charles L. 1977. *The Public Use of Private Interest*. Washington, DC: Brookings Institution.

Schuman, Howard, and Stanley Presser. 1981. *Questions and Answers in Attitude Surveys: Experiments on Question Form, Wording, and Context*. New York: Academic Press.

Schuman, Howard, and Cheryl Rieger. 1992. Historical Analogies, Generational Effects, and Attitudes toward War. *American Sociological Review*, 57: 315–326.

Schwartz, Joe, and Thomas Miller. 1991. The Earth's Best Friends. *American Demographics* (February): 26–35.

Schwepker, Charles H., and T. Bettina Cornwell. 1991. An Examination of Ecologically Concerned Consumers and Their Intentions to Purchase Ecologically Packaged Products. *Journal of Public Policy and Marketing*, 10(2): 77–101.

Scott, David, and Fern K. Willits. 1994. Environmental Attitudes and Behavior: A Pennsylvania Survey. *Environment and Behavior*, 26: 239–260.

Sears, David O., and Leonie Huddy. 1987. Bilingual Education: Symbolic Meaning and Support among Non-Hispanics. Paper presented at the Annual Meeting of the American Political Science Association. Chicago, IL.

Sears, David O., Richard R. Lau, Tom R. Tyler, and Harris M. Allen Jr. 1980. Self-Interest vs. Symbolic Politics in Policy Attitudes and Presidential Voting. *American Political Science Review*, 74: 670–684.

Seuss, Dr. 1971. *The Lorax*. New York: Random House.

Shabecoff, Philip. 1992. Shades of Green in the Presidential Campaign. *Issues in Science and Technology*, 9: 73–79.

Shabecoff, Philip. 2000. *Earth Rising: American Environmentalism in the Twenty-first Century*. Washington, DC: Island Press.

Shipan, Charles R., and William R. Lowry. 2001. Environmental Policy and Party Divergence in Congress. *Political Research Quarterly*, 54(3): 245–263.

Shutkin, William A. 2000. *The Land That Could Be: Environmentalism and Democracy in the Twenty-first Century*. Cambridge, MA: MIT Press.

Silver, Brian D., Barbara A. Anderson, and Paul R. Abramson. 1986. Who Overreports Voting? *American Political Science Review*, 80: 613–624.

Sims, Rodman A. 1993. Positive Attitudes Won't Make Cash Register Ring. *Marketing News* (June 7): 4.

Sinclair, Patti K. 1992. *E for Environment: An Annotated Bibliography of Children's Books with Environmental Themes*. New Providence, NJ: Bowker.

Singhapakdi, Anusorn, and Michael S. LaTour. 1991. The Link between Social Responsibility Orientation, Motive Appeals, and Voting Intention: A Case of an Anti-Littering Campaign. *Journal of Public Policy and Marketing*, 10: 118–129.

Skelton, George. 1986. Deukmejian Opposes Three Controversial Propositions. *Los Angeles Times*, September 3, 3.

Slovic, Paul. 1987. Perception of Risk. *Science* (April 17): 280–285.

Slovic, Paul, Baruch Fischhoff, and Sarah Lichtenstein. 1980. Facts and Fears: Understanding Perceived Risk. In Richard C. Schwing and Walter A. Albers Jr., eds., *Societal Risk Assessment: How Safe Is Safe Enough?* (181–214). New York: Plenum Press.

Smith, Kevin B. 1994. Abortion Attitudes and Vote Choice in the 1984 and 1988 Presidential Elections. *American Politics Quarterly*, 22: 354–369.

Smith, Tom W. 1978. Age and Social Change: An Analysis of the Association between Age-Cohorts and Attitude Change, 1972–1977. *GSS Social Change Report No. 5*. Chicago: National Opinion Research Center.

Smith, Tom W. 1979. A Compendium of Trends on General Social Survey Questions. *NORC Technical Report 15*. Chicago: National Opinion Research Center.

Smith, Tom W. 1984. Cycles of Reform? A Summary of Trends since World War II. Paper presented to the American Sociological Association, San Antonio, August.

Smith, Tom W. 1990. Liberal and Conservative Trends in the United States since World War II. *Public Opinion Quarterly*, 54: 479–507.

Smith, Tom W. 1994. Is There Real Opinion Change? *GSS Social Change Report*, No. 36. Chicago: National Opinion Research Center. ⟨http://www.icpsr.umich.edu/GSS/rnd1998/reports/s-reports/soc36.htm⟩.

Smith, V. Kerry. 1990. Can We Measure the Economic Value of Environmental Amenities? *Southern Economic Journal*, 56: 865–878.

Smith, V. Kerry. 1995. Does Education Induce People to Improve the Environment? *Journal of Policy Analysis and Management*, 14: 599–604.

Solomon, Lawrence S., Donald Tomaskovic-Devey, and Barbara J. Risman. 1989. The Gender Gap and Nuclear Power: Attitudes in a Politicized Environment. *Sex Roles*, 21: 401–414.

Soskis, Benjamin. 2000. Green with Envy: Hooray for Those Vacuous Celebrities. *New Republic* (May 8): 13.

Soto, Lucy. 1997. Many Favor an Increase in Gas Prices. *Atlanta Journal and Constitution*, November 21, A21.

Spangler, M. B. 1980. *Methods, Problems, and Issues in Environmental Cost Benefit Analysis of Nuclear Power Plant Alternatives.* Environmental Impact Assessment Course, Argonne National Laboratory for the International Atomic Energy Association, Illinois.

Stanfield, Rochelle L. 1988. It's Hip to Be Clean. *National Journal*, 20 (June 4): 1510.

St. Clair, Jeffrey. 1997. The Twilight of "Gang Green." *In These Times*, 21: 14.

Sterngold, Arthur, Rex H. Warland, and Robert O. Herrmann. 1994. Do Surveys Overstate Public Concerns? *Public Opinion Quarterly*, 58: 255–263.

Stimson, James. 1989. *Public Opinion in America: Moods, Cycles, and Swings.* Boulder: Westview Press.

Stisser, Peter. 1994. A Deeper Shade of Green. *American Demographics* (March): 24–29.

Stokes, Donald E. 1963. Spatial Models of Party Competition. *American Political Science Review*, 57: 368–377.

Streitwieser, Mary L. 1997. Using the Pollution Abatement Costs and Expenditures Micro Data for Descriptive and Analytic Research. *Journal of Economic and Social Management*, 23: 1–25.

Sudman, Seymour, and Norma M. Bradburn. 1982. *Asking Questions.* San Francisco: Jossey-Bass.

Sullivan, John L., James Pierson, and George E. Marcus. 1982. *Political Tolerance and American Democracy.* Chicago: University of Chicago Press.

Sunstein, Cass R. 1990. Remaking Regulation. *American Prospect*, 3: 73–82.

Sunstein, Cass R. 1997. *Free Markets and Social Justice.* New York: Oxford University Press.

Swan, James A. 1971. Environmental Education: One Approach to Resolving the Environmental Crisis. *Environment and Behavior*, 3: 223–229.

Switzer, Jacqueline Vaughn. 1997. *Green Backlash: The History and Politics of Environmental Opposition in the U.S.* Boulder: Rienner.

Taylor, Bron. 1997. Environmental Values in American Culture (book review). *The Ecologist*, 27 (January–February): 38–39.

Taylor, D. Garth. 1980. Procedures for Evaluating Trends in Public Opinion. *Public Opinion Quarterly*, 44: 86–100.

Taylor, Jerry. 1992. Campaign Trail Littered with Environmental Wrecks. *Plain Dealer*, December 5, 7b.

Teles, Steven. 1996. Think Local, Act Local. *New Stateman*, 126 (August 22): 28.

Tognacci, Louis N., Russell H. Weigel, Marwin F. Wideen, and David T. A. Vernon. 1972. Environmental Quality: How Universal Is Public Concern? *Environment and Behavior*, 4: 73–86.

Traugott, Michael W., and Paul J. Lavrakas. 2000. *The Voter's Guide to Election Polls* (2nd ed.). New York: Chatham House.

Tucker, William. 1982. *Progress and Privilege: America in the Age of Environmentalism.* Garden City, NJ: Anchor/Doubleday.

Van Liere, Kent D., and Riley E. Dunlap. 1980. The Social Bases of Environmental Concern: A Review of Hypotheses, Explanations, and Empirical Evidence. *Public Opinion Quarterly,* 44: 181–197.

Van Liere, Kent D., and Riley E. Dunlap. 1981. Environmental Concern: Does It Make a Difference How It's Measured? *Environment and Behavior,* 13: 651–676.

Verba, Sidney, and N. H. Nie. 1972. *Participation in America: Political Democracy and Social Equality.* New York: Harper & Row.

Vig, Norman J. 1994. Presidential Leadership and the Environment: From Reagan and Bush to Clinton. In Norman J. Vig and Michael E. Kraft, eds., *Environmental Policy in the 1990s* (71–95). Washington, DC: CQ Press.

Vig, Norman J., and Michael E. Kraft, eds. 1994. *Environmental Policy in the 1990s.* Washington, DC: CQ Press.

Wall, Glenda. 1995. Barriers to Individual Environmental Action: The Influence of Attitudes and Social Experiences. *Canadian Review of Sociology and Anthropology,* 32: 465–489.

Wasik, John. 1992. Market Is Confusing, But Patience Will Pay Off. *Marketing News* (October 12): 16.

Weigel, Russell, and Joan Weigel. 1978. Environmental Concern: The Development of a Measure. *Environment and Behavior,* 10: 3–15.

Westholm, Anders. 1997. Distance versus Direction: The Illusory Defeat of the Proximity Theory of Electoral Choice. *American Political Science Review,* 91: 865–883.

Wiley, John P., Jr. 1998. Coming to Terms. *Smithsonian* (December): 28.

Williams, Rolla. 1987. Window to the Wild. *San Diego Union-Tribune,* June 4, B2.

Wilson, James Q. 2000. Why They Don't Campaign about the Environment. *Slate* (October 27). ⟨http://slate.msn.com/?id=92005⟩.

Wilson, James Q., and Edward Banfield. 1964. Public Regardingness as a Value Premise in Voting Behavior. *American Political Science Review,* 4: 876–887.

Wilson, James Q., and Edward Banfield. 1965. Voting Behavior on Municipal Public Expenditures: A Study in Rationality and Self-Interest. In Julius Margolis, ed., *The Public Economy of Urban Communities* (74–91). Baltimore: Johns Hopkins University Press.

Winski, Joseph M. 1991. Big Prizes, but No Easy Answers. *Advertising Age* (October 28): GR-3.

Wirthlin Worldwide. Various years. *National Quorum* [datafile]. 1996; August 21–23, 1997, $n = 1,040$; September 11–14, 1998, $n = 1,010$; 1999. McLean, VA.

Wlezien, Christopher. 1995. The Public as Thermostat: Dynamics of Preferences for Spending. *American Journal of Political Science*, 39: 981.

Worldwide Concern about the Environment. 1989. *Our Planet* 1, no. 2/3: 14–15.

World Wildlife Fund (WWF). 1997a. State of the Climate (USMELL.092997) [datafile]. August 11–14, $n = 800$. Administered by the Mellman Group. Washington, DC: WWF.

World Wildlife Fund (WWF). 1997b. Voters Believe Global Warming Is a Reality. ⟨http://www.panda.org/climate_event/poll.htm⟩. Washington, DC: WWF.

Yankelovich Clancy Shulman. 1992. Survey [datafile]. January 16, $n = 1,000$. Westport, CT: Yankelovich Clancy Shulman.

Zaller, John. 1992. *Report on 1991 Pilot Items on Environment*. Ann Arbor, MI: National Election Studies (February 2).

Zaller, John, and Stanley Feldman. 1992. A Simple Theory of the Survey Response: Answering Questions versus Revealing Preferences. *American Political Science Review*, 36: 580.

Zisk, Betty H. 1987. *Money, Media, and the Grassroots: State Ballot Issues and the Electoral Process*. Newbury Park, CA: Sage.

Index